Ladybirds ○ /

The Untold Story of Women Pilots in America

by

Henry M. Holden

With Captain Lori Griffith

For Harriet, Jackie and Amelia;
you are the force.

First Edition
First Printing - Sept. 1991
Second Printing - March, 1992, revised

Cataloging in Publication Data

Holden, Henry M.

Ladybirds: the untold story of women pilots in America/ by Henry M. Holden with Captain Lori Griffith. --
p. cm.
includes index
ISBN 1-87963-011-7

Women in aeronautics. 2. Air Pilots -- History. I.
Griffith, Lori. II Title.

TL521   629.130 92   -
                          MARC

This book is available for purchase by educational institutions and groups at bulk discount prices. Write the publisher for information.

Table of Contents

Preface

Introduction

Acknowledgements

About the Author

About Lori Griffith

Foreword

Preface

In 1987, while I was in Long Beach, California, researching my first book, *The Douglas DC-3*, I decided to take a day off and visit the San Diego Aerospace Museum. I found an excellent display of aircraft and history. I found something else. In one section of the museum was a large display on the history of American women in aviation set up by a local chapter of the Ninety-Nines, the organization of women pilots formed by Amelia Earhart in 1929. The size of the exhibit surprised me. It was easily ten times the size of our National Air and Space Museum's exhibit on women in aviation.

The exhibit enchanted me. I browsed for over an hour learning more about American women and their role in aviation. What I discovered was a series of aviation milestones and successes that women in general and our country in particular should be proud of. I also learned the downside, that although there were qualified women pilots like Ruth Law and the Stinson sisters they were prohibited from flying in the military during World War I.

Women were an integral part of the Barnstorming era, and into the 1930s, when so many aviation advances took place. During World War II, they flew every fighter and bomber type manufactured in the United States. Of course they were still not allowed to fly in the military or in combat but they did form a para-military organization and flew as ferry pilots. It is from one of the Women Airforce Service Pilot (WASP) photos that I derived the name for this book. (See page 89.)

Women have played a major role in aviation's development and yet I had never read much about their individual accomplishments (except for women like Amelia Earhart and Jackie Cochran).

I began to search for reading material on this topic and found very little. Of course, books had been written that addressed women in aviation in general, usually grouping American women and European, but most of this material was at least a decade out-of-date. The Smithsonian Institution had three excellent monographs on American women and there were also about two dozen magazine articles but nothing on contemporary American women in aviation. The pioneering efforts of many male pilots are not common knowledge either; however, documentation on male contributions is much easier to find than the female contributions. There was virtually no material that a young woman could turn to for a role model and I wondered why. Without indicting generations of innocent Americans the answers I soon discovered were attitude and economics. These factors are, in some cases, but to a lesser extent, still operating today.

In 1973, small changes began to take place. The first two women were hired by the airlines and the Navy opened flight training to women. A decade later, the number of women flying for the airlines had grown to just over 200. Almost twenty years later there are over 1,200 women flying for American commercial airlines and about 600 military pilots. The figure is still small but the trend can not be halted or reversed.

Today there are over 50,000 women licensed as pilots in the United States. There has been notable progress in the last decade, and I see more progress in the 1990s than in the entire 80 years women have been involved in aviation. Over the years, progress for the most part has been made quietly and always with dignity. These women have given a new meaning to the words courage, and determination and with few exceptions have been ignored, forgotten or never acknowledged.

One day in the future, when there is wide-spread acceptance of women pilots and acknowledgement of their accomplishments, Harriet Quimby, Amelia Earhart, and Jackie Cochran will smile to each other and say, "What's next, ladies?"

Harriet Quimby

INTRODUCTION

Men do not believe us capable, because we are women. Seldom are we trusted to do an efficient job. Now and then women should do for themselves what men have already done, and occasionally what men have not done, thereby establishing themselves as persons, and perhaps encouraging other women toward greater independence of thought and action.

Amelia Earhart

Many of us remember the first American in space, and the first man on the moon, but do you remember the first American woman to fly into space? Time may dull our memories, but most of us remember John Glenn, Neil Armstrong, and Sally Ride. We also recognize the name Amelia Earhart, who was prominent in early aviation, but how many of us recognize the names Blanche Stuart Scott, Harriet Quimby, or Bessie Coleman, all women who had "firsts" in early aviation? Even many people in aviation don't recognize the first woman to solo an airplane, the first to earn her pilot's license, or the first black woman licensed as a pilot.

America did not recognize the true value of these early air pioneers' accomplishments because women were excluded from those activities judged to be the province of men. The achievements today's women have in aviation are due to these early air pioneers, many of whom gave their lives in the pursuit of their goal.

During the two decades following World War I, the field of aviation grew by leaps and bounds. Women slowly assimilated into aviation. They set records only to be broken within weeks, or days, by other women. Then the women went after the men's records, often upsetting the old records. Air races became popular, aviation clubs and associations formed, oceans were crossed, transcontinental flights became common, and barnstormers and movie stunt pilots performed seemingly impossible feats of daring. Aviators went farther, faster, and higher than ever before, and women were unmistakably a part of these accomplishments. The women who followed the men into the air proved that they could compete with men, and fly just as well if not better. They endured the same hardships and should have earned the same recognition.

Some women like Amelia Earhart amd Jaqueline Cochran were highly visible and continually earned kudos and criticism, but most women in early aviation posed an economic threat to the men. Sometimes their success made them heroines for a short while, but their failures were used to prove they were physically and emotionally unfit for flying. Ironically, if they survived an accident, it was used to show that air travel was safe.

Influential persons in aviation were aware of women's efforts and accomplishments and could have helped to expand the roles of women in aviation, but they were surprisingly restrictive in their views. Eddie Rickenbacker took the executives of Boeing to task in 1930 for hiring the first female flight attendants. He argued that flying was a man's occupation and should stay that way. Ironically, Ellen Church, the first flight attendant, was a pilot and was seeking employment as such when Boeing hired her to serve food, and look after passengers.

This book focuses on many "first" events but it recognizes all women, on the ground and in the air, who contributed as role models, in large and small ways to the science and advancement of aeronautics. The late Judith Resnick said, "Firsts are only the means to the end of full equality, not the end itself."

Judith Resnick

Acknowledgements

It is difficult for me to adequately acknowledge and thank all the people who made this book possible. I have listed the contributors on page 208. If I have omitted anyone it is by oversight.

There are several people who deserve special mention. Lori Griffith deservingly shares the cover with me because over the years she has believed in this project and has been a great supporter. It was Lori who first gave me access to the ISA through their newsletter. The inputs from the ISA allowed me to gain insight and a perspective known to few outside the industry. Lori flew up from Charlotte, N.C., to Newark, New Jersey, on her day off to sit with me and discuss the details of the book. She read the entire manuscript for technical accuracy, and statements that may have been unclear, misleading or incorrect. Her feedback and suggestions were of immeasurable help. Without her this book would not have been possible.

Jean Ross Howard, chairman of the board of the Whirly-Girls provided me with much of the material that appears in Chapter 5. She generously read early drafts and gave me excellent feedback. She invited me to a "hovering" where I met some of the Whirly-Girls in this book. Jean made phone calls and wrote letters on my behalf that resulted in an enthusiastic input from the Whirly-Girls. I must deservingly credit her with making Chapter 5 not only possible, but an integral part of this history.

WASP historian Marty Wyall also responded generously to my call for help. She provided me with access to what I believe is probably every photo ever taken of the WASPs. There were literally hundreds. I am only sorry I could not use more of them.

My research assistant, Carole Sandhovel, again came through as she always does, spending untold hours on drafts and redrafts, made countless phone calls, and follow-ups, in addition to juggling other responsibilities.

My special thanks to Pamela McFadden who generously volunteered her time to read the manuscript and give me a woman's perspective on this work. She also edited the first printing for typos and in doing so helped make this story better.

My son Stephen also provided me with invaluable help by reading at least three versions of this manuscript. His gentle but constructive comments were invaluable to me.

And my son Scott, and wife Nancy, thank you for giving me your support and space I needed to write this book.

After all the help and feedback, I continued "tweaking" the manuscript, so any errors appearing in this book are the result of my own attempts at fine tuning this work, and may not reflect the hard work and suggestions of my many friends.

To everyone, my thanks, best wishes, and blue skies.

Jacqueline Cochran

About the Author:

This is Henry M. Holden's fourth book. His previous books are, *The Douglas DC-3; The Boeing 247 - The World's First Modern Commercial Airplane;* and *The Fabulous Ford Tri-Motors.* He has published more than 200 magazine and newspaper articles that have appeared in a New Jersey newspaper syndicate, and the *New York Times Syndication.* His articles have appeared in various national magazines including, *USAir, Peterson's Photographic* , and various aviation magazines including the prestigious American Aviation Historical Society's *Journal,* and *Airport Press.* He is a member of the Aviation Hall of Fame of New Jersey, the American Aviation Historical Society, The International Group for Historic Aircraft Recovery (TIGHAR), and the Aviation/Space Writers Association. The author lectures frequently on aviation topics and is currently working on his next book, *The Whirly-Girls.*

About Captain Lori Griffith

Captain Lori Griffith received her Private Pilot Rating when she was 18, and attended Indiana State University on a Talent Grant Scholarship in their aviation program. She began in 1979 and finished a four year degree with a double major in Aviation Administration and Professional Flight in two years.

Immediately following College she was hired by a company in Kentucky where she flew pipe line patrol of oil and gas lines. In August of 1981, she was hired by Atlantis Airlines a Commuter in the Southeast. Because She was so young she could not receive her ATP until her 23rd birthday. A captain's position came available in early December, 1983, and both the Chief Pilot and Training Officer flew the line for her to hold the slot until her birthday. On that day, she was the youngest female Captain in the industry.

She was hired by Piedmont Airlines in March, 1984, as a Flight Engineer on the Boeing 727. One year later she up-graded to First Officer, and then in March of 1987, received her Captains bid on the Fokker F-28.

Captain Griffith has the following ratings; Commercial, Instrument, Multi-Engine, Seaplane, Rotocraft, Glider, Instructor Airplane, Instructor Seaplane, Instructor-Instrument and Instructor-Multi-Engine. She is presently working on her Balloon Rating, and dreams of completing a "Blimp" or Airship Rating.

Today, Captain Griffith is a Boeing 737 captain with USAir. Captain Griffith is also a past Executive Council member of the International Society of Women Airline Pilots (ISA + 21) as well as their Master Seniority List Chairperson. She is a member of the International Whirly-Girls, the Wings Club, and the Ninety-Nines.

Foreword

As a female airline captain flying in the 90s, I stand in a unique position on the line of women aviator progression. I respect the pioneers who flew before me for giving their lives and paving the way for the opportunities my generation enjoys in the present. At the same time I can look ahead in confidence to the positive role that women will play in aviation's future which is yet to unfold.

Whether we measure our progress by social barriers or sound barriers, women have made their mark in aviation. They did so by meeting challenges, making sacrifices and resisting social and economic pressures all for the dream to fly.

We are still on the road to winning acceptance and becoming a more visible force in the aviation industry, but acceptance comes with time. Soon there will be a day when a woman's presence in the cockpit isn't questioned and when all female flights cease being a novelty.

As you read the individual stories of these women pilots and their struggles and sacrifices you will find an underlying theme too glaring to ignore. Flying is a driving force that knows no gender, a power so awesome that handsome salaries and public notoriety only become secondary benefits, and the yearning to fly is so strong that any sacrifices are justified in the end.

Women who enter the aviation field today need only worry about challenging themselves. There are hundreds of doors that have already been opened and numerous "firsts" yet to be scored and all to the up and coming women who dare to reach for the sky!

Captain Lori Griffith

CHAPTER 1

The Early Years: 1910-1919

The aeroplane should open a fruitful occupation for women. I see no reason they cannot realize handsome incomes by carrying passengers between adjacent towns, from parcel delivery, taking photographs or conducting schools of flying.

Harriet Quimby - June 1912

Blanche Stuart Scott

It was Friday, September 2, 1910, and the morning sun was quickly burning away the ground fog at Curtiss Field, on Long Island. The day promised to be sunny and warm. Blanche Stuart Scott was busy checking the wires and bolts on a single-engine Curtiss "Pusher" airplane.(Fig. 1-1)

Scott was thoroughly familiar with the construction of the machine. Glenn Curtiss, the machine's builder, had reluctantly given Scott instruction in the care and maintenance of one of his prize machines. Curtiss was also a clever business man. Scott insisted on learning to fly, but Curtiss believed aviation was the exclusive province of men. Giving Scott lessons meant money for his fledgling aircraft company. On the other hand, if she crashed the machine it would mean a tarnished reputation for his company, and show poor judgement on his part for allowing a woman to fly. People too might blame his fragile machine, and not the ineptness of a woman. Curtiss decided to give her lessons since there were other options also available to him.

The day started out as others, with no particular attention to the only woman student pilot in the entire country. Most people knew Curtiss was too smart to actually allow her to fly. Scott had driven the machine back and forth across the runway often, and occasionally it would leap a few feet off the ground. It would always come right back down, like a rabbit hopping across a field. None of the machine's spastic hops even resembled sustained flight. Curtiss was in his office at the time, and had seen Scott by the rickety machine. As in the past, he paid little attention to her. Scott, on the other hand, was intently curious. Nothing she did to the machine could get it into the air. On this day she was poking around the engine, when she found something strange. It was a small piece of wood wedged beneath the throttle lever. The wood seemed to retard the throttle's full motion. Scott smiled to herself and removed the piece of wood. She climbed into the machine and watched the mechanic crank the propeller to prime the engine. A moment later the engine coughed white smoke and came to life.

Scott proceeded to taxi the machine across the grass and line the craft up between two white sticks marking the runway. At this point the sound

Fig. 1-1 Blanche Stuart Scott

of the engine caught Curtiss' attention. It seemed louder than usual. He peered over his spectacles, and out the window. What he saw made him drop his pencil. The frail, fabric-covered wooden machine was roaring down the makeshift runway, gathering speed. Suddenly it was airborne. Curtiss rose from his seat staring in disbelief. The craft was about forty feet off the ground, and gaining altitude.

Curtiss' jaw fell open. He moved somnolently toward the window, slowly removing his spectacles. His only thoughts, "She's flying -- she'll crash -- she'll ruin me."

Scott felt the wind in her face and a rush of exhilaration. The gust of wind that lifted the machine from the invisible ties to earth had died away. She felt the plane sinking. She moved the throttle an inch forward and the sinking stopped. Enough for today, she thought. She eased the frail machine back down to the green grass and rolled to a stop.

Blanche Stuart Scott became the first American woman to solo in a fixed-wing, heavier-than-air machine. She was, however, no stranger to being the first at something. She had also been the first woman to drive an automobile across the United States. In those days, there were less than 300 miles of paved roads in the country.[1]

The people quickly gathered around Scott, asking dozens of questions. She looked at Curtiss and simply said, "Something must have happened to the throttle block." She was grinning as the words left her lips. She was the first and last pupil Curtiss ever taught himself. People soon discovered he could not deal very well with women, especially independent women.

That brief flight convinced Scott that she wanted to be a pilot. She felt flying was in her blood. She teamed up with various exhibition groups and toured the country for about six years. One of her favorite tricks was flying upside down under bridges. Another, and probably her best trick was the "Death Dive," a stunt in which she hurtled down from 4,000 feet, leveling off barely 200 feet above the ground.

Blanche Scott was the first American woman pilot to solo but not the first American woman to go aloft. That distinction goes to Miss Lucretia Bradley, of Phillipsburg, N.J., who flew in a balloon in 1854.

Scott had several accidents, some serious. One accident grounded her for more than a year. During this long recuperation she decided flying was becoming too risky. "There seemed to be no place for a woman engineer, mechanic or flier," she said. "Too often people paid money to see me risk my neck, more as a freak, a woman freak pilot, than as a skilled flier."[2] Scott retired from aviation in 1916, at the age of twenty-seven. Her decision was almost unique. The early aviation years exacted a horrendous toll on pilots, and many of her friends died in crashes. Scott later considered herself lucky and privileged to participate in the beginning of aviation and see an American land on the moon. (Ms. Scott lived 81 years, folding her wings in January, 1970. The Post Office honored her with a stamp in 1988.

Bessica Raiche

Today there is a mild disagreement over which woman flew first. Since it has never definitely been established whether Scott's first flight was accidental or intentional, some give the credit of the "First Woman Aviator of America" to Bessica Raiche, for her flight on October 13, 1910.

Raiche never thought there was any rivalry between them. "Blanche deserved the recognition, but I got more attention because of my lifestyle. I drove an automobile, was active in sports like shooting and swimming, and I even wore riding pants and knickers. People who didn't know me or understand me looked down on this behavior. I was an accomplished musician, painter and linguist, I enjoyed life, and just wanted to be myself."

---

[1] *The History of the 99s.* The 99s Inc. Publishers; 1979 p.6

[2] Valerie Moolman. *Women Aloft.* Va. Time Life Books, 1981; p.8

Raiche became interested in aviation when she visited France and saw the Wright brothers flying their airplane. When she and her husband Francois returned to America, they built their first airplane. It was a Wright-type craft which they innovated with bamboo, piano wire and silk. They built the pieces in the living room of their home, and assembled it in the backyard.

Without any prior flight instruction, Raiche successfully made her first solo flight. Later she admitted her actual flying instruction consisted of a mechanic placing the wheel in her hands saying, "Pull it this way to go up, and push that way to go down."[3]

Raiche and her husband formed the French-American Aeroplane Company and began building lightweight airplanes. They were the first to substitute silk for the heavy fabric covering, and piano wire for the commonly used iron stove wire.

Raiche flew regularly and when she became seriously ill, her doctor said the air in the clouds was bad for her lungs. She took his advice, retired from aviation, and moved west. Once settled, she decided to enter medical school. After she received her degree in medicine, she became a practicing physician, and put the fun of flying behind her.[4]

Harriet Quimby

Harriet Quimby was one of the most popular figures in early aviation. She was born in 1875, the youngest of two daughters to William and Ursula(Cook) in Coldwater, Michigan. William Quimby was a Union Army cook, a one-time farmer and grocery store owner. Harriet received her primary education in the local public schools. [5]

She continued her education after the family moved to San Francisco when Harriet was entering her teens. Throughout her formative years her mother's influence molded her character. Her mother's brother, a doctor, was renowned for herbal medicine cures. Harriet and her mother sat at home bottling these curatives while her father sold them from a wagon. Harriet spent many hours at her mother's elbow learning the ways of late 19th century San Francisco. The proceeds from the medicines enabled Harriet to attend the better schools in the big city. Harriet had one goal; to be a journalist. At the age of 21, she found work on a newspaper. A year later she landed a job with the popular *Leslie's Weekly*, as their drama critic, moved to New York, and her life took a new turn.

Quimby saw her first airplane when she was 35 years old. She had gone to Belmont Park, on Long Island, searching for a dramatic story when she witnessed John Moisant fly from Belmont Park to the Statue of Liberty and back, 36 miles in 34 minutes. He won the Statue of Liberty Race and the event ignited a fire for flying in Harriet's heart.

3
   The History of the 99s. The 99s Inc. Publishers; 1979 p.6
4
   Claudia Oaks. *United States Women in Aviation Through World* War I.
   Smithsonian Studies in Air and Space #2. 1979, p.19
5
   Congressional Record 6/4/87

"There were not many flying schools that would accept women," said Matilde Moisant, Harriet's close friend. "Goodness, they didn't want us driving motor cars, much less airplanes." Moisant had a good connection. She and Quimby enrolled in her brother's flying school on Long Island. Because of the social pressures, the two women dressed disguised as men for their lessons. Harriet always took her lessons at sunrise. At that time of day the lessons didn't interfere with her work, the air was usually calm, and she could keep her activities a secret, or so she thought. When a reporter discovered Harriet's charade she was happy to see it end. The newspapers gave her a lot of publicity. The press described Harriet as a "willowy brunette" and they quickly tagged her with the nickname the "Dresden China Aviatrix," because of her "beauty, daintiness, and haunting blue-green eyes." Others, however, sensationalized. One headline shouted, "Woman in Trousers, A Daring Air Pilot." Quimby became almost notorious, but she was a strong woman. "The notoriety bounced off her," said Matilde Moisant. "Harriet enjoyed the publicity."

In the infant days of aviation the license requirements were to fly five alternate right and left turns around pylons, and complete five figure-eights. Harriet Quimby completed this part of the test with ease, but the second part required her to land within 100 feet of where the plane had left the ground. On her first attempt she landed too far from the spot. The following day a crowd, attracted by the fast-spreading rumors that a woman had attempted to earn a pilot's license, watched as Quimby successfully repeating all the previous tests, and came within seven feet, nine inches of the mark, setting a record. (There were no foot brakes in those days.)

Quimby became the first American woman to earn her pilot's license on August 1, 1911. Her instruction covered 33 lessons with a little more than four-and-one half hours in the air. Harriet Quimby was the second woman in the world to earn her license. (The first was Raymonde de la Roche of France, in 1910.)

"After the flight," Quimby said, "I climbed out of my monoplane and nonchalantly walked over to the official observer and said, "Well I guess I get my license."

"I guess you do," he said reluctantly.[6]

After the flight, a *New York Times* reporter covering the event described Quimby as having "her face covered with grease and dirt, and her blue-green eyes flashing happily."

"Flying seems easier than voting," quipped the beaming Harriet to a group of women looking on.

"I took up flying," Quimby told reporters, "because I thought I'd like the sensation, and I haven't regretted it. I like motoring but after seeing monoplanes in the air, I couldn't resist the challenge. The airlanes have neither speed laws or traffic policemen, and one need not go all the way

[6] *New York Times* August 2, 1912

around Central Park to get across to Times Square. Then too, it's good to be the first American woman to earn a pilot's license." (Fig. 1-2) Not long after Quimby earned her license she set the first record. On September 4, 1911, in front of a crowd of 15,000 persons gathered at the Richmond County Fair, she became the first woman to make a night flight. [7]

Matilde Moisant received her license just 12 days later,and wasted no time getting started. She won the Rodman Wanamaker Trophy for a

Fig. 1-2
Harriet
Quimby

[7]   *Staten Islander* Sept. 7, 1911

woman's altitude record reported to be 2,500 feet; the first woman awarded an altitude trophy and a truly remarkable accomplishment in those days, given the instability of the aircraft. [8]

Matilde Moisant was a bit of a maverick, and admonished by the sheriff of Nassau County, on Long Island, for flying on Sunday. When she didn't stop he decided he would curtail Moisant's activities by arresting her. She avoided him by flying to another airfield. Later a court decided that flying on Sunday was no more immoral than driving a car on that day. Moisant was also a firm believer in the "lucky" number 13. She was born Friday, September 13th, 1887, and she was the second woman to qualify for her pilot's license on the 13th day of August, 1911. (Her license reads the 17*th*.) She always numbered her airplanes *13*.[9]

"Matilde and I parted company for awhile," said Quimby. "She went on to fly with Moisant International Aviation, touring the South and Mexico. When John died in an air accident, her enthusiasm for flying began to taper off." Matilde Moisant also had three serious accidents which also dampened her enthusiasm.

Moisant picked April 13, 1912 to announce she would make her last flight the following day. It turned out to be nearly the last thing she ever did. Her aircraft developed a fuel tank leak as she came in for a landing. By the time she landed, the plane and her clothes were on fire. Moisant owed her escape to the heavy tweed knickers, coat, and thick leather helmet she was wearing. True to her word, she gave up flying. (Matilde Moisant lived longer than most of her contemporaries, male and female. She folded her wings in Los Angeles, in 1964, at the age of 77. (Fig. 1-3)

Harriet Quimby was goaded from time-to-time by the press. On one occasion, right after she received her license they asked her why she limited her flights to short hops. Quimby replied confidently, "I have not found it very difficult to control my monoplane in short flights and there is no reason I soon shouldn't make longer ones." Soon Quimby was making 20 mile trips.

In 1912, flying was considered dangerous, especially for the "weaker sex." Because of this attitude, Quimby had difficulty with public opinion. Since there were few great American male aviators at the time, the result was that most Americans looked on aviation as a dangerous sport. Since Quimby was the highest profile female flyer they still doubted her ability even after she proved she could fly.

Quimby's response to these critics was, "There is no more risk in an airplane than in a high speed automobile, and a great deal more fun. More American women drive cars than English or French women, and

[8]
Katherine Brooks Pazmany. *United States Women in Aviation 1919-1929.* Smithsonian Studies in Air & Space #5; p.32

[9]
Claudia Oaks. *United States Women in Aviation Through World War I.* Smithsonian Studies In Air and Space #2; p.31

Fig. 1-3
Matilde
Moisant in
her heavy
tweed flying
suit

yet there are already several French women aviators. Why shouldn't we have some good women air pilots, too?" [10]

Another problem arose as women began to fly. People wondered what to call a female air pilot. Some editors used the word aviatrix, and others airwoman. Still another said why discriminate? Why not lump all pilots under one term, aeronauts?

The *New York Times* said, "The difficulties to be overcome in establishing a permanent glossary of terms for this rapidly developing science of aeronautics seem formidable." The *New York Times* finally reported that the neutral term "aviator" was acceptable for both sexes. America, however, was not ready for a non-gender related term, and *aviatrix* remained in use for many years.

"Once I had my license," said Quimby, "I realized I could share the thrills of aviation with my readers. I wrote in the first person because they would feel closer to the events in the cockpit. I called some of my

[10]   *New York Times, May 11, 1912*

adventures, 'How a Woman Learns To Fly,' and 'The Dangers of Flying and How To Avoid Them.'" Quimby also wrote a thrilling account of her English Channel crossing, "An American Girl's Daring Exploit." Harriet tried to show American women that there were alternatives that would give them new adventurous directions in which to chart their futures.

Quimby was the first woman to pilot a plane across the English Channel, and the second woman to cross it by air. On April 2, 1912, Eleanor Trehawk-Davis flew as a passenger across the Channel in a plane piloted by Quimby's friend, Gustav Hamel. Two weeks later, on April 16, Quimby borrowed his new Bleriot monoplane for the trip.

"On the eve of my Channel crossing, my friend Hamel was convinced that no woman could make the trip alone. He was so anxious for my safety and ability to pilot the airplane across the Channel, he suggested he dress up in my flying costume, fly across the Channel and land at a remote spot where I would be waiting to take the credit. I adamantly refused his offer, and if he were not such a dear friend I would have been very angry. I did accept his offer to help me read a compass. That was something new to me."

On Sunday, April 14, Harriet went to the Aerodrome three miles outside Dover. She had never flown the plane and wondered about its control. She had learned to fly in a 30 hp. plane and this plane was 50 hp. The weather was perfect and she could make out Calais, 22 miles across the Channel. Everyone urged her to take off immediately and take advantage of the weather. She refused. She would not fly on Sunday, for any reason.

Monday brought thick clouds and heavy rain, and Harriet and her small ground crew sat all day in a cramped automobile, waiting for the weather to clear.

On Tuesday, the rain had stopped but there was heavy fog in the area. Nevertheless, Harriet decided it was time. At 5:39 a.m. Quimby took off and aimed her plane for France.

"It took me about 30 seconds to reach an altitude of 1,500 feet. As I looked down, Dover Castle was in a veil of mist. I could barely see the tugboat filled with reporters sent out by the *Mirror* to follow my course. The fog quickly surrounded me like a cold, wet, gray blanket."

Hamel had only given Quimby brief training in reading a compass. She had never seen one operating in a moving plane. "I had never used a compass, and I was doubtful of my ability to read one. Hamel said it would be shaking from the engine vibration. I was hardly out of sight of the cheering crowd before I hit a heavy fog bank, and found the compass to be of invaluable help. I could not see above, below ,or ahead. I started climbing to gain altitude, hoping to escape the fog. It was bitter cold, the kind that chills to the bone."

Under her flying suit of wool-backed satin, Quimby wore two pairs of silk combinations, a long woolen coat, over this an American raincoat, and around her shoulders a long wide stole of sealskin. Even this did not satisfy her solicitous friends. At the last minute they handed her a large hot water bag, which Hamel insisted on tying to her waist, like an enormous locket.

"I did not suffer from cold while crossing. The excitement, I guess stimulated me. I noticed when I landed the hot water bag was ice cold. I'm sure it helped, but I didn't notice."

During the flight Harriet recalled Hamel's remark about the North Sea. If she drifted off course by as little as five miles, she would get lost and probably go down in the icy waters. (Hamel should have taken his own advice as seriously as he had given it to Quimby. A few years later he flew off into the Channel mist and never returned.)

Quimby followed the compass needle faithfully. It was her only frame of reference and reassurance. The fog was beginning to take its toll on her nerves and she descended, looking for clear air. As she lost altitude the machine tilted to a steep angle, causing the gasoline to flood the engine. The machine began to backfire. She had but one option, she thought. "I figured on pancaking down to strike the water with the plane in a floating position. To my great relief, the gasoline quickly burned away and my engine began an even purr. I glanced at my watch and estimated I should be near the French coast." [11]Soon a gleaming strip of white sand, flanked by green grass, caught her eye and she knew she was safe. Because of winds and the underpowered 70 hp. engine, the 22-mile trip took one hour and nine minutes.

The editorial page of the *New York Times* on April 18, 1912, took a narrow view of Miss Quimby's accomplishment. The editorial was no doubt influenced by the paper's lack of support for the women's suffrage movement that was in full bloom in the spring of 1912. The *Titanic* ocean liner had also just been lost. "Even when so much public attention is on the loss of the *Titanic,* that a woman alone, and depending wholly on her own strength, skill, and courage, has driven an aeroplane across the English Channel, does not pass unnoticed.

"Miss Quimby's flight is a considerable achievement. Just a few months ago this same flight was one of the most daring and in every way remarkable deeds accomplished by man. Since then the passage has been repeated by men, and now for them there is little or no glory. The flight is now hardly anything more than proof of ordinary professional competency."

The *Times* continued and in a condescending warning said, "The feminists should be somewhat cautious about exulting over Miss Quimby's exploit. They should not call it a great achievement, lest by so doing they invite the dreadful humiliating qualification, great for a woman.

"A thing done first is one thing; done for the seventh or eighth time is different," they said. "Of course it still proves ability and capacity, but it does not prove equality."

The smell of these sour grapes still lingered by the time Harriet arrived back in New York, on May 12. She received no hero's welcome,

11
Harriet Quimby. "An American Girl's Daring Exploits." *Good Housekeeping,* September, 1912

and there was no ticker-tape parade. It was a matter of timing. Only a week earlier, 15,000 women and 619 brave men marched up Fifth Avenue in support of women's suffrage. The male leaders of the city had not yet recovered from this demonstration of feminine assertiveness. They weren't ready to admit there were female eagles, let alone honor them. In an ironic note, the suffragists also gave Harriet the cold shoulder; to them the only thing worse than an anti-suffragist was an independent woman uninterested in their cause.

Harriet was not a woman who would let some anonymous editor have the last word. "I wish I could express my views on this," she said. "It's not a fad, and I did not want to be the first American woman to fly just to make myself conspicuous. I just want to be first, that's all, and I am honestly delighted. I have written so much about other people, you can't imagine how much I enjoy sitting back and reading about myself for once. I think that's excusable in me."

Today we can put Harriet Quimby's accomplishment in proper perspective. Without any of the modern instruments, in a plane that was hardly more than a winged skeleton with a motor, and one with which she was totally unfamiliar, to fly across the 22-mile English Channel, in 1912, required extraordinary courage, skill, and self-confidence.

Harriet was just as honest about her other motivation for flying; money. She was already a successful journalist with a weekly article and theater section in *Leslie's Weekly*. She wanted a good income to support herself and her parents, but she also had a special goal. She had decided that she wanted to be financially independent so she could devote full time to her first love, creative writing. At the time, only about 25% of the female population worked outside the home, and their annual income was just about $500.[12]

Harriet Quimby had many vocal critics concerned about her "immodest dress" Her critics said she would corrupt the public's morals. A major problem women aviators faced in those early days was what to wear. For centuries society severely restricted woman's dress and fashions. Skirts dragged the floor and corsets contoured the female figure into grotesque "S" curves, and exaggerated hourglass shapes. Frills, ruffles, and lace also weighed down the costumes and gave women the look of over-decorated dolls. Long flowing skirts that barely showed a booted ankle had just become fashionable in 1910. (Fig. 1-4)(In 1915 fashions had not changed much for women. A Massachusetts school department manual dictated that women, "Must not wear any dress more than two inches above the ankle.) [13] Wide hats were in fashion, but they were impossible to wear in an open airplane. Pants or trousers were the most practical form of clothing for pilots. For most American women emerging from the Victorian era, however, pants were unacceptable or immodest. Some women attempted compromise and wore trousers with

---

[12] Nancy Zerfoss, "Schoolmarm to School Ms." *Changing Education*, Summer 1974, p.23

[13] ibid

Fig. 1-4
Harriet
Quimby (r)
& Matilde
Moisant.

rows of buttons on the inside that converted the garment to a skirt when not flying. Most women found this uncomfortable, awkward, and sometimes dangerous. Eventually a flying outfit evolved for women pilots that gained public acceptance. The modest attire was a two-piece outfit, a blouse and wide-legged tweed knickers or riding pants with high-top boots and a soft fabric helmet with goggles. Harriet approached her critics and this challenge with an air of dignity, flamboyance, and style. "It may seem remarkable," she said, "but when I began to fly I could not find a regular aviator's suit that would fit me in New York. Finally my tailor helped me design a suit that I hoped would establish a standard for the proper flying costume for women in this country." The outfit was extraordinary for 1912, a one-piece purple satin outfit with full knickers reaching below the knee, and high laced black kid boots below. Her head gear resembled a monk's hood, and her accessories were flying goggles, elbow-length matching gauntlet-style driving gloves, and a long leather coat for cold weather flying. In warmer weather she wore a full length cape to match the purple outfit. Quimby had pioneered the "jumpsuit."

Harriet Quimby was very happy at the Boston Air Meet of July 1, 1912. She was at the peak of her career, and receiving rave accolades wherever she went. Everyone loved her, and she would soon be financially independent. Quimby had never had a flying accident because she was a careful and capable pilot. She gave close attention to every wire and fastener before each flight.

"I had confidence in my craft," she said. She had just written an article for *Good Housekeeping* entitled, "Aviation As A Feminine Sport." Its aim was to give women confidence in their ability to equal the performance of men. "There is nothing to fear if one is careful," she said. "Only a cautious person should fly. I never mount my machine until I check every wire and screw. I have never had an accident in the air. It may be luck, but it is also to the care of a good mechanic."

At 5:30 p.m., it was Harriet's turn to fly her routine. Harriet was going to break the over-water speed record of 58 miles per hour. Again she was flying an unfamiliar plane. She had brought the plane back with her from France, and in an ironic twist, after the boat had docked and she had cleared Customs she was called back into the inspector's office. No one knew how to classify her Bleriot monoplane. After a long discussion, Customs officials put it under the category reserved for polo ponies. "One flying machine - equivalent to 70 horses."

With a passenger, William P. Willard, the manager of the event, Harriet took off over the 27-mile course to the Boston Light. As she came out of a turn around the lighthouse at an altitude of 5,000 feet, the plane turned over sharply and nosed down toward the bay. As the horrified crowd watched, first Willard's body, and then Quimby's fell from the plane. The bodies tumbled through the air and plunged into the harbor waters. Quimby died on impact, and Willard drowned. Ironically, the Bleriot monoplane flew itself out of the dive, and glided into the water. It nosed over on impact but was not damaged.[14]

Blanche Stuart Scott, who was airborne and witnessed the tragedy, was barely able to land her plane. "With an effort plainly visible from the earth," the *Times* reported, "she (Scott) turned the nose of her machine downward, came to a landing like a flash, and fainted before anyone could reach her."[15]

Before her death, Harriet had written an article giving her opinion of aviation as a career for women. The article, for *Good Housekeeping* Magazine, did not appear until September, two months after her fatal crash. "I think," she said, "the aeroplane should open a fruitful occupation for women. I see no reason they cannot realize handsome incomes by carrying passengers between adjacent towns, from parcel delivery, taking photographs or conducting schools of flying. Any of these things it is now possible to do. The number of men fliers will always outnumber the women, just as chauffeurs outnumber the women who drive automobiles. If we establish fuel supply and landing stations, there is no reason we cannot have airlines for distances of 50 to 60 miles. This mode of travel

14 Claudia Oaks. *United States Women in Aviation Through World War I.* Smithsonian Studies in Air and Space #2; 1979.

15 Valerie Moolman. *Women Aloft,*. Va. Time Life Books, 1981; p.24

would be particularly delightful during the Summer, allowing one to escape the heat and dust that make overland travel so uncomfortable."

The editor added a preface to Miss Quimby's article, "In view of her tragic death, there is a note of pathos in the enthusiasm and in the prophecy for women fliers in her article."

The *New York Sun* also commented on her tragic death. "Miss Quimby is the fifth woman in the world killed while operating an aeroplane (three were students) and their number thus far is five too many. The sport is not one for which woman are physically qualified. As a rule they lack strength and presence of mind and the courage to excel as aviators. It is essentially a man's sport and past time." While the maelstrom of criticism whirled around Harriet's tragic death, Miss Quimby was quietly buried in Kenisco Cemetery, in Valhalla, New York.

More than 80 years later, time has vindicated Harriet Quimby. Her spirit at the time was no doubt angered by the editorials. If she were here today she would smile and rejoice in saying, "See, I told you so." (On April 27, 1991, the Post Office issued a stamp honoring Miss Quimby. She is the third woman aviator to be honored, Blanche Scott and Amelia Earhart are the others.)

Julia Clark

Julia Clark was the third woman to earn a pilot's license, on May 19, 1912. Julia was born in London, immigrated to the U.S., became a citizen, married a westerner, and settled in Denver. Julia enrolled at the Curtiss Flying School at North Island in San Diego, and like Scott, soloed in a Curtiss plane and then joined an exhibition team. On June 17, 1912, she decided to make a test flight around dusk. Visibility was poor, and on takeoff, one wing struck a tree limb. The plane tumbled into the ground pinning her beneath the wreckage. She was the first American woman to die in an air accident, and her death preceded Quimby's by two weeks.[16]

Why Fly?

You may wonder why Harriet, Julia and other women gambled their lives in such unstable machines. There are two reasons: First, they didn't know how a stable aircraft should act. There was no experience, either male or female, in designing stable and aerodynamic aircraft. Amelia Earhart later gave a reason for women past and future. Before Earhart's last flight, she wrote to her husband, "I want to do it because I want to do it. Women must try to do things as men have tried. When they fail, their failure must be but a challenge to others."[17]

There was very little scientific engineering in the early aircraft, with most engineered by ear and by eye. There were few instruments, and little consideration for safety. Often a pilot maneuvered through turns

16
Claudia Oaks. *United States Women in Aviation Through WW I.* Smithsonian Studies in Air and Space # 2. 1979 p. 20

17
Amelia Earhart. *For the Fun Of It* Introduction N.Y. Harcourt Brace, 1932

with the help of a long cloth tied on the upper wing of a biplane. If the cloth did not blow straight in a turn, the pilot knew the craft was slipping too much.

The propeller was driven by a chain (probably requisitioned from the Wright brothers bicycle shop), that sometimes broke. When this happened the chain would snap guy wires and shred wing fabric. The wing warping device would sometimes flex the already strained airframe too far with disastrous results. Most of the planes in this first decade of aviation were dangerously unstable. The Bleriot had killed a dozen pilots, mostly in Europe, and its instability was not well known at the time of Quimby's death. The early Bleriot monoplanes needed the weight of two people (or sand bags in place of a passenger). If the passenger's weight shifted, as first thought in Quimby's accident, the plane sometimes became uncontrollable.

It was also dangerous to fly these airplanes unless wind and weather conditions were nearly perfect. The low horsepower motors did not lend much control to the fragile craft.

Marjorie Stinson, an early aviation pioneer wrote she was lucky if she flew five minutes in the morning and five in the evening during her training because the wind created such danger. During the day she said, she spent most of the time fishing and flying kites.

It was not unusual in those days for aviators not to use safety belts, or parachutes. The low powered planes did not generate much centrifugal force in turns. Most pilots refused to wear a safety belt fearing that in a crash the weight of the plane's heavy, often rear-mounted engine would crush them (like Julia Clark). They were sure they would be able to dive free at the last moment and avoid such a fate. No one in the early days gave much thought to falling *out* of the airplane.

The deaths of Miss Quimby and Mr. Willard began to focus attention on the need for universal acceptance of seat belts on airplanes. Glenn Martin, an aircraft designer, felt Quimby's accident was avoidable. "Miss Quimby's accident would not have happened if they were wearing straps. When going at a terrific speed, the machine, on striking a hole in the air, will drop suddenly and lift one from his seat. I wear straps on my aeroplane and even then I am thrown violently up against them by such drops."[18] Lincoln Beachley, a well-known aviator, had for the day a typical analysis of the tragedy. "Miss Quimby was coming down from five thousand feet under full power. She was a light, delicate woman and it could easily have happened that the terrific rush of air was too much for her and she became weakened and unable to control her levers."[19]

Like a meteor, Harriet Quimby flashed through the skies of America, and her show was breathtaking in its brilliance. But as suddenly as a meteor appears, it is gone. Harriet Quimby was gone and the skies over America suddenly seemed darker with her passing. In her short time she

[18] *Boston Globe,* July 2, 1912
[19]   ibid

had left a legacy for women, but she did not survive the first anniversary of her license.

Miss Quimby's death became almost mystical and her superstitious critics have blamed it on her jewelry. From the start of her flying career, she was the most flamboyant personality in aviation circles at the time. She, like her contemporary male pilots, was superstitious. One way of maintaining her femininity was to wear her lucky jewelry on every flight. The jewelry has become a legend, the legend of her misfortune.

She once owned a bracelet and necklace fashioned from the tusk of a wild boar. The tusk at one time was part of an East Indian idol called *Genasha*. The legend said it brought bad luck to its former owner, a French pilot. Before Harriet bought this idol, people say her life was happy, well ordered, and filled with good fortune. When she began to wear the jewelry, they say her luck changed. She had disappointments, misfortune, and finally disaster. Eleanor Trehawk-Davis flew the English Channel as a passenger just before Harriet, and died in a later air accident. Trehawk's death reinforced the legend when people discovered Harriet had loaned the "lucky" necklace to Mrs. Davis. After Harriet's death, the jewelry disappeared. The story goes that her mother wanted to donate the pieces to the Smithsonian Institution but they disappeared from the body. So, the legend developed, that the thief would have ill fortune. The jewelry remains lost and perhaps still brings its errant owners bad luck.

Katherine & Marjorie Stinson

On July 24, 1912, Katherine Stinson became the fourth and youngest (age 16) woman in the United States to earn a pilot's license. Unlike many of her contemporaries, she had a rewarding career in aviation. Her younger brothers, Eddie and Jack, also became prominent in aviation. Jack was an early aviator, and Eddie, first a test pilot, later founded the Stinson Aircraft Company. Her younger sister Marjorie also became a prominent aviator. In 1913, Katherine Stinson participated in the New Years Day Pasadena Parade, in California. She flew along the parade route with her rose-adorned plane. A small aviation milestone for sure, but she was the first pilot and plane to participate in a parade.

Katherine became an instant success and between 1913 and 1914, she went on the tour circuit. In August 1914, her sister Marjorie graduated from Wright flying school and they appeared together in Kansas City.

"Our mom (Emma Stinson) was a real angel," said Katherine. "The encouragement she gave us to look to the sky is primarily responsible for our success. She allowed me to fly at the tender age of sixteen, and signed a release that Wilbur Wright required before he would enroll my sister Marjorie in his flying school."[20]

[20]
Claudia Oaks. *United States Women in Aviation Through World War I.*
Smithsonian Studies In Air and Space #2. 1979, p.36

The family moved to San Antonio, Texas, and started a flying school. Emma Stinson put up the money needed to open the school, and she became her daughters' business manager. Katherine and Marjorie began to give the local boys flying lessons, and taught Jack and Eddie to fly. By 1916, their school was thriving. They had fourteen students.

When America entered World War I, Marjorie and her sister tried to enlist as military pilots. America was still not ready for women to be more than nurses, and rejected them. Undaunted, Marjorie and her sister then opened a school to train American and Canadian men as pilots for the war. The Stinsons did have supporters, however few. New York Congressman Murray Hulbert introduced a bill in Congress to permit women to join the Flying Corps and go to France; however, this bill did not pass.

"I didn't feel I was doing enough for the war effort," said Katherine, "so I went to France, and became an ambulance driver." While Katherine was at the front she contracted influenza, and returned to the United States. Katherine's health deteriorated, forcing her to retire from aviation in 1920.

Marjorie also had a great career while it lasted. In May, 1915, at 17, she became the first woman authorized to fly the experimental airmail service. It was unofficial and little mention is found in history books. The Post Office did not restart airmail service until 1918. (Fig. 1-5)

Even more prominence came to young Katherine Stinson on July 18 1915, at Cicero Field, in Chicago. She became the first woman to loop-the-loop. A male pilot followed and repeated the loop. She would

Fig. 1-5 Marjorie (l) & Katherine Stinson

not allow a man to outdo her so she expanded on her original feat. She developed her own special loop called the "dippy twist," a vertical bank in which the airplane rolls wing-over-wing at the top of the loop. She first performed this on November 21, 1915, and the next day outdid herself when she combined this maneuver with eight consecutive loops, flying

upside down for 40 seconds, and executing a series of spiraling spins. [21]
Katherine's most spectacular feat occurred on December 17, 1915, when
she attached railroad flares to her aircraft, a dangerous act in a wood
and fabric covered plane. She traced the letters "C A L " in the nighttime
sky, then looped, flew upside down, and spiraled down to within 100 feet
of the ground trailing a shower of sparks.

Katherine Stinson went on to set many endurance records, and an
early non stop, cross country distance record for both men and women.
On a Liberty Loan Drive she raised more than $2 million. Cross country
flying was not easy in those days. She used a topographical map between
Buffalo and Albany, then followed the New York Central Railroad tracks
to New York City. She continued past New York City using a Pennsylvania Railroad map to find her way to Washington, D.C.

Cross country flying in those days was also dangerous. There were
practically no maintenance facilities where emergency repairs could be
made. It was customary also to ship airplanes used at air shows by
railroad. To put it simply, even to some of the pilots, airplanes were not
a safe or a serious means of transportation.

Obstacles

There were a variety of obstacles to women's progress in aviation. As
aircraft design became more scientific, so did the balance between
aircraft size, weight, and maneuverability. With this balance, the physical
strength barriers that prevented smaller women from flying were no
longer restrictive. The size and strength of the pilot became less important, and more women climbed into the cockpit.

Aviation was a male-dominated field and there were strong op-
ponents to women's progress. Claude Grahame-White, a famous British
aviator, confessed that he had reluctantly taught some women to fly. His
ambivalence came from wanting to make more money teaching people
to fly and betraying the narrowly restricted fraternity. Flying was a rich
man's sport and those men who did fly found maintenance costs a
constant drain on their resources. Teaching flying would offset some of
these costs. Graham-White felt he had pushed women toward an early
death. He did not doubt their courage, he said condescendingly, but
feared what he called an innate lack of self confidence. In an emergency
he said they would, "panic and lose control." He also felt they were
"temperamentally unfit, and when calamity overtakes them, and sooner
or later it will," he said, he would feel, "in a way responsible for their
sudden decease."[22]

There were a few men who believed that women had a place in
aviation, but sometimes they did more harm than good. In 1911, Professor Rudolph Hesingmuller, of Vienna, published a list of reasons he
thought women made better pilots than men:

21
   ibid
22   1911 Newspaper clipping, Matilde Moisant biographical file, NASM

Because she has retained the primitive facility of seeing with full retina; enforced modesty and flirting have caused this;
Because she has scattered attention instead of concentration; this is invaluable to an aviator who must notice many things at once;
 Because she has the facility of intuition, that quality of the mind that can take in many things simultaneously and induce a conclusion, an essential in aviation;
 Because her specific gravity is less than a man's;
 Because she needs less oxygen and can better meet the suffocating rush of air;
 Altitude effects her less than it does a man;
 Because her sneezes, in a man an actual spasm, are controlled by ages of polite regression;
 *Because she feels more quickly warming atmospheric changes;
 Because she loves speed.

Most people recognized that the professor's reasons were ludicrous, but his simplistic ignorance cast the seeds of doubt. In some circles, his reasons were used for the reverse, to illustrate why women would fail.

World War I
By the time the United States entered World War I, women had become, albeit in small numbers, permanent participants in aviation. The 1917 Aero Club of America Bulletin listed 11 licensed women pilots. There were many more women who were competent aviators but because of social pressures, they had never bothered to take the license test.

Many of the unlicensed women flew for recreation, and others just to meet the challenge. The automobile was gaining widespread usage and some women who had driven the new motor cars, like Harriet Quimby and Bessica Raiche, saw the airplane as an extension of the motor car adventure. Some women, like some men, found flying required more skill and dexterity than the automobile, and soon quit. Others viewed aviation as a new means of getting around. Most women who flew saw aviation as a way of establishing themselves as individuals.

Women could not serve their country in the military as pilots but that did not stop them from contributing to the war effort. Others used their flying skills at Red Cross and Liberty Loan drives, and by doing so, helped focus more attention on women in aviation. Bernetta Miller, Alys McKey Bryant, and Helen Hodge found other ways to serve. Miller joined the Women's Overseas Service league and went to the front as a canteen worker. She was awarded the Croix de Guerre and numerous American citations for her work. Bryant submitted various applications to fly in combat but ended up as a test pilot and instructor. For a time she assisted the Goodyear Company in building military dirigibles. Hodge taught U.S. aviation cadets and made exhibition flights for the war effort. [23]

From their unselfish efforts these women earned from some men genuine but distant respect. Opinions about women pilots ranged from a small and strange minority to trouble makers trying to upset the status quo. Hidden behind almost every objection voiced by men was this last fear. An upset in the male aviation community would have grave economic and social consequences. Women would steal jobs and gain power. "Look what was happening to society as a result of the sufferage movement!" some said. Women were voicing their opinion! What if women were welcomed into aviation?

The Industrial Revolution introduced women to the workplace, first in the factories. In America they were the backbone of labor, first in textiles, but soon they were branching out to manufacturing. Not until World War I did opinions change about their ability to help in the manufacture of airplanes. It was ironic too, because women had already proven capable of flying the machines.

During World War I the shortage of men at home forced the airplane factories to open their doors to women. Industry leaders only agreed to allow women into the airplane factories because of the biased assumption that women had a "natural ability" to work with fabrics and a particular aptitude for sewing. Of course, the first jobs women had were sewing fabric for the wings, and cutting and sewing parachute cloth (men still packed the parachutes).

Women quickly proved themselves capable workers. The manpower shortage was so critical that factories then tested the women's skills further. They got jobs involving light machine work, acetylene welding, and radiator core making. On the light machine work they came close to both the men's output and quality. On inspection jobs where the work involved handling many small parts their work far excelled that of men. Yet for equal or better performance of identical labor, they were paid far less than the men, generally about 40 per cent less . By the end of the war women had made significant, albeit temporary progress. About 25,000 women worked in the aircraft industry. After the war, all forward motion for women in the direction of progress came to a stop. Most women returned to the home, freeing up the jobs for the returning male veterans. But there had been progress.

Ground Support

Someone once said that behind every great man stands a woman. In aviation this was certainly true. Behind the scenes there were dozens of women motivating, both emotionally and financially some of the successful men. They were not pilots but their contributions to the advancement of aeronautics are also no less significant. Many women who helped men

---

[23] Claudia Oaks. *United States Women in Aviation Through World War I.* Smithsonian Studies in Air and Space #2

conquer the air and who remained on the ground supporting them are now forgotten. Some have not received the credit they deserve, either from history or the men they helped, and some had all the reward they wanted in the association with the men they helped and loved. A few have modestly refused to acknowledge their important role.

Katherine Wright, sister of Wilbur and Orville, contributed to their scientific pool of knowledge and to their bank account throughout their struggle to conquer flight. Almost every historian credits her with using the money she earned teaching Latin and Greek to purchase the materials for their fragile airplanes. One even says she mortgaged her home when Wilbur was negotiating for the sale of the manufacturing rights, and Orville needed to join him in France. Orville often spoke of her support before their first historic flight and in later years, when she helped them establish their airplane business. "When the world speaks of the Wrights," he said, "it must include my sister; she inspired much of our effort." [24]

On the day Orville crashed at Fort Meyer, Virginia, Katherine Wright hurried from Dayton to spend weeks at his bedside while he recovered. She also became a passenger on their flights when the brothers went to Europe. She was there primarily to reassure people that flying was safe. Many people still wonder why she never learned to fly. She did not marry until 1926, and then tragically she died of pneumonia in 1929.

One woman, hidden in the shadow of her great husband, did not seem to mind. Mrs. Alexander Graham Bell had poor hearing and inspired many of her husband's inventions in the transmission of sound. She stood beside him during the lawsuits for patent infringement brought by other alleged inventors of the telephone. She stood beside him, too, while he worked with men forty years younger, trying to solve the riddles of mechanical flight. Mrs Bell sold some of her property to get money to organize the "Aerial Experiment Association." This group served to carry out her sixty year old husband's interests in aviation. It was $25 a week for expenses from this group that coaxed Glenn Curtiss out of his motor shop and into aeronautics.

Another woman who contributed to aviation in a different way was Miss E. Lillian Todd, of New York City. Miss Todd, one of the first women to design and build an airplane was neither an engineer nor a pilot. She was a stenographer. At the December, 1906 meeting of the Aeronautical Club of America, she exhibited an aircraft she had designed herself. Without an engine the craft was no more than a glider. In theory the pilot would run down an incline and the rushing air would cause two fans to revolve and catch the air. The trapped air would start two propellers and the plane would theoretically fly. Once flying, the incoming air would continue revolving the propellers, keeping the craft aloft. Reducing the air flow to the fans through shutters would slow the craft and allow the pilot to land. This sounded like a perpetual motion engine, and did not

[24] ibid p.7

take into account factors then unknown, like drag and friction. Miss Todd had no training in the use of tools or mechanics but built the craft entirely by herself. Her designs eventually drew attention from such notable figures as Andrew Carnegie and Harry Guggenheim. There are no records to show the craft ever flew, but her work did stimulate aviation. She went on to design and built several other full scale planes, but none of these ever flew. She founded the Junior Aero Club, and taught dozens of youngsters to build flyable model planes, and she was responsible for several pursuing aviation careers. Miss Todd continued to design planes and developed ideas that were later widely accepted, like folding wings and collapsible planes for easier storage and transportation.[25]

First Flighter

Another woman aviation proponent was not a pilot and not an aircraft designer, but the widow of a Pennsylvania tanner. This unsung hero went a long way to promote the idea of women flying, but in a different way. Mrs. Clara Adams became known as "The most persistent first flighter."

Mrs. Adams was no stranger to flying machines and piled up an impressive record. Her first flight was in a Thomas Airboat in 1914. By 1924, she was the guest of Dr. Hugo Eckener, and the first woman on the first flight of the German Zeppelin ZR-3, "Los Angeles."

In 1928, she bought the first ticket sold to fly across the Atlantic on the Graf Zeppelin. She was the only woman passenger and the ticket cost $3,000. In 1931 when the giant 162 passenger Dornier "DoX" flying boat hopped the south Atlantic, she flew to Rio, got aboard, and flew into New York, the only woman paying passenger. In 1936, Mrs. Adams was aboard the Hindenberg on its maiden flight to America. Mrs. Adams was the first of eleven women to board.

She made the inaugural commercial passenger flight across the Pacific in 1936, and the following year made the New York to Bermuda inaugural flight of the Pan American Airways service. In 1931, she had reserved a ticket on the first Pan American Clipper to fly the Atlantic. Eight years later she finally used the reservation. Pan American Airways, delighted with her perseverance, gave her the royal treatment. She was also the first woman to fly around the world using the available regular passenger service. She beat the previous record set by a man. She made the trip in one day, 19 hours, and 52 minutes. The publicity she received from these flights encouraged women to try this new form of transportation. World War II cut her adventures short. The emergency curtailed the supply of civilian flights, and Mrs. Adams joined the rest of America's women in the war effort. [26]

---

[25]   ibid

[26]   Claudia Oaks. *United States Women In Aviation 1930-1939*. Smithsonian Studies In Air and Space #6. p.28

CHAPTER 2

Barnstorming 1920-1929

There is a world-old controversy that crops up whenever women attempt to enter a new field. Is a woman fit for that work? It would seem that a woman's success in any particular field would prove her fitness for that work, without regard to theories to the contrary.

Ruth Law

Ruth Law was at the Boston Air Meet and watched Harriet Quimby fall to her death. Just one month earlier she had enrolled in the Burgess Flying School, and took her first plane ride on July 1, 1912. Quimby's death had shaken Law but did not stop her. She had tasted a sense of fulfillment in flying and would not return to the traditional role defined by society. "I purchased a Wright biplane because it seemed to me they had the greatest success. Harriet died in a monoplane but that didn't scare me. I figured it was the monoplane's fault." [1]

On August 1, 1912, Ruth soloed, and received her license on November 12, 1912. She immediately went to work as a commercial pilot flying passengers to and from the Sea Breeze Hotel, in Florida. She soon was able to buy a Curtiss Pusher "Loop Model," and began flying aerobatics at Daytona Beach, Florida.

On November 19-20, 1915, Law made the greatest flight of her career, and set three new records. On a 590-mile non-stop flight from Chicago to New York, she broke the American non-stop cross-country record, the world's non-stop cross-country record for women, and the world's second best non-stop cross country record. [2]

It was also, she admitted, the most difficult flight she ever made. "I had to improvise a lot. I had to design a supplementary fuel system for the flight that increased the aircraft's auxiliary tank capacity from eight to fifty-three gallons. I also built a device so I could read maps without letting go of the controls." It was a simple map case holding a scroll of geodetic strip maps. She held the vertical control with her left hand, and held the horizontal control with a device fitted to her right knee. This let her turn the map scroll with her free right hand. (Fig. 2-1)

Law also had some tense moments. On the last leg of the flight, over Manhattan, her engine began to sputter. To reach her final destination,

---

[1] Betty Peckham. *Women in Aviation*. New York, Thomas Nelson & sons, Inc. 1945.

[2] Claudia Oaks. *United States Women In Aviation Through World War I*. Smithsonian Studies In Air and Space #2, p.39

Fig. 2-1 Ruth Law (r)

Governor's Island, Law had to bank the plane, to get the fuel from the tanks to the carburetor. When she landed they found the fuel tanks dry.

When the war broke out, the government would not allow Law to fly in combat. They did realize the publicity value if they allowed her to be the first woman to wear the uniform of a non-commissioned officer while participating in Army and Navy recruiting drives. Like the Stinson sisters, she flew exhibition flights to raise money for Red Cross and Liberty Loan drives. During one of these flights, she pushed the plane to its limits, and set a new woman's altitude record of 14,700 feet.

The 1920s - Aviation Grows

The general attitude that flying was socially inappropriate and physically impossible for all but the strongest women was still common as America entered the third decade of aviation.

The United States had emerged from the war years with a huge surplus of airplanes and male pilots. With this backdrop, there was not much chance of women taking to the air in large numbers. The war curtailed civilian flying and at the end of the war there were only a handful of qualified women pilots. Former Army pilots and aircraft mechanics flooded the market, all looking to use their new skills.

The glut of surplus aircraft forced prices down and airplanes became affordable for many people. With few jobs available in civilian aviation, many pilots bought their own planes and hit the barnstorming trail.

Law saw the pilot glut after the war, and went touring overseas to China, Japan, the Philippines, and Europe. When she returned in 1920, she and her husband formed a three-plane troupe called *Ruth Law's Flying Circus*. One of her most popular tricks was the plane-to-car transfer. To appease her fans her stunts had become increasingly dangerous. Law didn't see them as dangerous. "My trickiest one involved climbing out of the cockpit, and inching toward the center of the biplane's wing. The pilot would then make three loops with me standing on the wing."

Barnstorming was not as glamorous as it appeared to many people. Many men and women lived near the poverty level with maintenance and fuel costs eating up the little money they earned. Only dedicated pilots could tolerate the financial hardships and constant exposure to weather.

For women, the barnstorming circuit was the only open avenue to a career in aviation. Many women who wanted to fly began by barnstorming. A few flew passengers, but most raced, and set records. Some, like Law, did well financially.

"I was earning $9,000 a week at one point. Imagine, $9,000 a week in the 1920 economy. I think too, my success encouraged other women to stretch their wings."[3]

One morning in 1921, Law's flying career came to an abrupt, but fortunately not tragic end. As she sat at the breakfast table, she opened the newspaper and read where she had retired from aviation. When her husband saw the expression on her face, he confessed. Fearing for her life, he had sent the announcement to the newspapers without consulting her.

"Of course I was angry for awhile but I took a pragmatic view of aviation." Law was thinking about Laura Bromwell, who set a world's record on May 21, 1921, when she flew 199 consecutive loops before a crowd of 10,000. Laura stopped only when she ran out of fuel. She died two weeks later in a crash. Bromwell had wanted to be the first woman to fly across the United States, and she had not flown the craft she was intending to use. Friends told her not to try aerobatics in an unfamiliar plane. She went out of control completing a loop and crashed. Investigation later revealed that she did not wear her seat belt. She was small and her height had restricted her access to the rudder pedals. Her death caused an uproar.

Over the years the *New York Times* had inched toward the acceptance of women in aviation, or at least that they were on the scene permanently. In an editorial, it said, "So many men have lost their lives in airplane accidents that individual attention to the long list of names had ceased to cause any deep emotions except in the minds of their relatives and

[3] ibid, p.42

friends. When a woman is the victim, however, the feeling of pity and horror is as strong as that produced by the first of these disasters to men. Men did not abandon aviation because of its dangers, but the death of Laura raises in many minds at least the question if it would not be well to exclude women from the field of activity in which their presence seemed unnecessary."

The *Times* continued, "There was little if any reason for assuming that Laura as an aviator was less competent than a man of the same training and experience. She had made many flights safely, and though there seems to be some evidence of carelessness in her preparations, there was nothing characteristic of her sex. Nothing unlike many male aviators have done sometimes with like results. Laura was careless not to have her holding straps in perfect order. But that should not be counted against women as a class, to compete with men in this new profession."[4]

Law explained her pragmatism, "We all take risks, and so often do we go aloft when conditions are not perfect that the principle of safety first should keep us on the ground. Since Law did not need the money she deferred to her husband's wishes, and dropped out of aviation. Her parting words echoed her recognition of progress in aviation but not of women's progress. "Things had become so proper, so many rules and regulations. The good old crazy days of flying are gone."[5]

Nelle Zabel Willhite

Prerequisites for women pilots in the 1920s were courage, stamina, and the will to survive. Nelle Zabel Willhite not only had the resilience to endure, but extraordinary courage as well. Despite a bout of childhood measles that left her deaf, Nelle learned to fly. On January 13, 1928, she became South Dakota's first licensed woman pilot. Like many women in early aviation, she had a lonely life, but hers was extremely lonely. She was the only woman pilot for several thousand square miles. People avoided her not only because she was a woman invading a man's world, but because most people did not know how to deal with her deafness. Nelle became a regular and popular attraction at air shows and county fairs throughout the Midwest. She specialized in flour bombing and balloon racing. Balloon racing was both difficult and dangerous. For this event, pilots would try to fly into balloons floating in the air. It required a good deal of skill to make the sharp turns necessary to hit the balloons. During World War II, Nelle served as a ground school instructor and then as a propeller inspector on B-29 aircraft.

Exhibition flying required sturdy equipment, extraordinary skill, and determination of the pilot. Some women, like Amelia Earhart, did not

---

[4]   *New York Times.* June 7, 1921; p.16

[5]   Katherine Brooks Pazmany. *United States Women in Aviation 1919-1929.*
Smithsonian Studies in Air & Space #5; p.11

believe it pointed the way to progress but it showed the possibilities. As for women flying in exhibitions Earhart said, "It probably will be necessary for some time contrary to legal precedent that they (women) are considered guilty of incompetence until proved otherwise."[6]

Amelia Earhart, a woman 50 years ahead of her time, echoed Law's pragmatism. Many of her ideas were radical and to some inflammatory. Her position of being a world-wide celebrity gave her a global forum, but even with that large an audience, her views on women's role in society took many years after her death to become accepted.

The question came up constantly in the press and on the street. Why did women fly? Why were they not content with the structured life of a husband and family? The reasons were two-fold. First it may be in the dual nature of the word "flight" - in English the word implies both to fly and to flee. Second, flying took women someplace - away from their traditional and assigned roles and the dullness of routine existence. Margery Brown, a barnstormer and eloquent spokeswoman, answered the question this way: "Halfway between the earth and sky one seems to be closer to God. There is a peace of mind and heart, a satisfaction that walls cannot give. When I see an airplane flying I just ache all over to be up there. It isn't a fad, or a thrill or pride."

Brown believed flying would make women more independent, confident, and self-reliant, and she zeroed in on a reason all women understood. "Women are seeking freedom. Freedom in the skies! They are soaring above the temperamental tendencies of their sex that have kept them earth bound. Flying is a symbol of freedom from limitations."[7]

Bessie Coleman

In the 1920s the airplane was still new and held a curious fascination for thousands of Americans. Most pilots were men but increasing numbers of women could be found in the skies over America. For some women, barnstorming gave them nationwide attention.

A woman who fit this singular distinction was Bessie Coleman. She was born in Atlanta, Texas, on Jan. 26, 1893, the 12th of 13 children born to a former slave. Her mother motivated her and instilled a driving force that would help her fight overwhelming odds, blatant racism, and sexism to become the first licensed black American pilot. (Fig. 2-2)

From an early age her mother urged her to "become somebody." Although her mother could not read, she managed to borrow books from a traveling library, hoping that somehow Bessie could teach herself to read. "I did," said Coleman, "and I found a brand new world in the written word. I couldn't get enough. I wanted to learn so badly that I finished high school something very unusual for a black woman in those days. The teachers I had tried so hard. I don't wish to make it sound easy but I

---

[6] Amelia Earhart. *The Fun of It*. Harcourt Brace N.Y. 1932, p. 179

[7] Claudia Oaks. *United States Women in Aviation 1930-1939.* Smithsonian Studies in Air and Space

decided I wanted to go to college too. Since my mother could not afford college, I took in laundry and ironing, to save up the tuition money."

When Coleman thought she had enough saved, she enrolled in Langston Industrial College (now Langston University, Oklahoma). She had seriously underestimated the expenses though, and her money lasted only one semester. When she realized she couldn't go on she became depressed, moved to Chicago to live with her older brother, and found work as a manicurist.

"I guess it was the newspapers reporting on the air war in Europe during World War I that got me interested in flying. I was an avid reader, and searched the libraries looking for information on flying. I think all the articles I read finally convinced me I should be up there flying, and not just reading about it, so I started searching for a flying school. At first I thought it would be easy, just walk in and sign up. I didn't realize that I had two strikes against me. I remember hearing of a few women pilots before the war but I had never seen one. The other strike against me was my color. No one had ever heard of a black woman pilot in 1919. There

Fig. 2-2 Bessie Coleman

were few women pilots of course, but a black woman? I refused to take no for an answer. My mother's words always gave me strength to overcome obstacles. I knew someone important and decided to see if he could use his influence to get me into a flying school."

Robert S. Abbott, the founder and editor of the *Chicago Weekly Defender*, was very enthusiastic about Coleman's idea, but also pessimistic. After an exhaustive search of the flying schools in the country, he concluded that there were some who would teach a woman, but there were none that would teach a black woman. "He did have a ray of hope, said Coleman. "He told me that Europe had more liberal attitudes toward women and people of color and suggested I study French." At first Coleman couldn't believe it. If she wanted to fly she would have to travel across an ocean for her lessons.

Coleman took Abbott's advice and went to night school. In a few months she had learned enough French and saved enough money to travel to Europe. She underestimated how much it would cost, and her money ran out. She came back to America and found a job in a chili restaurant, but she would not let go of her dream. She would earn her pilot's license.

Coleman went back to Europe again in 1921. This time she had plenty of money and went looking for the best instructor she could find. Coleman learned to fly with the world famous aircraft manufacturer Tony Fokker's chief pilot. Fokker said she had a skill and what he called a natural ability. He encouraged her and since there were not many women pilots he was anxious for her to succeed. Coleman earned her license on June 15, 1921. She had made her dream come true.

"I returned to the United States with my air-pilot license from the Federation Aeronautique International. I was the first black pilot in the world. I had grand dreams but I was a realist. If I could have a minimum of my desires, I would have no regrets. Coleman looked on her achievement without significance unless she could share it with others. Having reached her first goal, she set a new one. "I decided blacks should not have to experience the difficulties I had faced, so I decided to open a flying school and teach other black women to fly. I needed money for this so I began giving flying exhibitions and lecturing on aviation. The color of my skin, at first a drawback, now drew large crowds wherever I went. At first I was a curiosity, but soon the public discovered I could really fly. Then they came to see "Brave Bessie," as they called me."

When Coleman went back to Texas, she ran into an age old problem. At one of her exhibitions the officials refused to let the blacks in the same entrance as the whites. "I wasn't going to let them humiliate *my* people, who were coming to see me. I told them I would not fly until they let the blacks through the same gate as the whites." The officials yielded to her demand but still separated the blacks inside. She didn't have enough clout to force that issue.

Like many of the early aviators, Coleman had several accidents. Her first occurred in 1924, in California, while doing an advertisement for the Firestone Rubber Company. The accident did not stop her: flying was in her blood and she continued giving airshows. She was attracting national attention. Nothing would stop her, not even discouragement from her friends and family. Even after witnessing the death of a student pilot, and herself suffering a broken leg and several broken ribs in a crash she would not weaken.

On April 30, 1926, with almost enough money saved to open her school, she had another accident. This time it was fatal. Bessie was performing in a May Day exhibition in Orlando, Florida, for the Negro Welfare League, of Jacksonville. It was 7:30 p.m. when Bessie, accompanied by her mechanic and publicity agent, William Wills, took her plane up for a test flight. Wills had taken the plane up on a test flight a week earlier and had landed twice because of engine trouble. [8]

Coleman had been in the air barely ten minutes and was at an altitude of 5,000 feet when she put the plane into a 110 mile per hour power dive. The plane suddenly flipped over, and Bessie, who had neither fastened her seat belt nor worn a parachute, was thrown from the plane and plunged to her death. Wills, trapped in the plane, died upon impact. Minutes after the crash a bystander lit a cigarette and unthinkingly tossed the lighted match to the ground, igniting the spilled gasoline. The wreckage went up in flames. No one knows why Bessie did not fasten her seat belt, or had not worn a parachute, but a later investigation found a wrench jamming her controls.

Was the misplaced wrench the fault of a careless William Wills? Some have suggested more than an accident. Bessie Coleman was an articulate black woman who had a dream for her people, 60 years ago, and therefore a threat.

Her friends returned Coleman's body to Chicago, the city she loved. They buried her in Lincoln Cemetery, on the city's southwest side.

On the 10th anniversary of her death, Abbott wrote an editorial in the *Chicago Weekly Defender*. "Though with the crashing of the plane life ceased for Bessie Coleman, she inspired enough members of her race by her courage to carry on in aviation and what they accomplish will stand as a memorial to Miss Coleman."

Although her death was attributed to pilot error, it might easily have been caused by equipment failure. Planes were expensive to maintain, and most women lacked access to a major source of funds, aircraft manufacturers. The aircraft manufacturers hired mostly male pilots to demonstrate their products at air shows and races. Many companies were reluctant to sponsor women, some because they didn't believe in women flying and others because they feared consumer backlash. So women usually flew with inferior or poorly maintained equipment.

"Woman are often penalized by publicity for their every mishap, said Amelia Earhart later. "Any disproportionate "breaks" they get when they are accomplishing something are lost in crash headlines. The result is that such emphasis sometimes directly effects chance for a flying job. I

---

[8]   *Ebony*. "They Take to the Skies." May 1981; p.88

had one manufacturer tell me that he couldn't risk hiring women pilots because of the way accidents, even minor ones, became headlines in the newspapers."[9]

Willa B. Brown

It would be a decade before another black woman would gain as notable a position in aviation as Bessie Coleman. That woman was Willa Brown. She was not the flamboyant barnstormer that Bessie Coleman was but her impact on aviation was significant. Brown held a Masters Degree from Northwestern University, a Master Mechanic Certificate, a commercial pilot's rating, and a CAA ground school instructor's rating. Her most notable achievement was leading the fight to have black men included in the Army Air Force in 1939.

She established the Coffey School of Aeronautics, the first black-owned private flight school in the United States approved by the U.S. government. Bessie Coleman's dream had finally been born. This school was responsible for the initial training of the men who became pilots in the 99th Pursuit Squadron the highly decorated all-black fighter squadron of World War II.[10]

Brown, also organized Squadron 613 of the Civil Air Patrol, and was the first black to hold the rank of officer in that organization.

First Attempt To Match Lindbergh

Charles Lindbergh's successful crossing of the Atlantic in May, 1927, gave many women the inspiration to answer the challenge of the skies. Women saw aviation as a chance to prove they were just as resourceful and daring as men. On August 16, 1927, spurred by Lindbergh's success, Mildred Doran became the first American woman to attempt a solo ocean crossing. James Dole, president of the Hawaiian Pineapple Company, sponsored a non-stop flight from Oakland, California, to Honolulu, and offered $25,000 as first prize. Since there had been two successful flights from the Hawaiian islands, everyone thought the contest would be a cinch. Three pilots died enroute to the race, and five more during the race. Two more were lost during the later search. Mildred Doran, one of the five lost at sea, was the first woman to lose her life in an ocean crossing. Although she was an experienced pilot, she had never gotten her pilot's license.[11]

Ruth Elder

Ruth Elder was a twenty-three year old some-time actress when she heard of "Lucky Lindy's" flight from New York, to Paris. She made up

[9] Amelia Earhart. *The Fun Of It.* Harcourt Brace N.Y. 1932, p.179

[10] Charles Paul May. *Women in Aeronautics.* Thomas Nelson & Sons, Inc. N.Y. 1962; p.145

[11] Kathleen Brooks-Pazmany. *United States Women In Aviation 1919-1929.* Smithsonian Studies In Air and Space #5; p20.

her mind that she would be the first Lady Lindy, the first woman to fly across the Atlantic. Her stage critics and others immediately held her in ridicule when she made her announcement. Some called her proposed flight a publicity stunt, prompted by Lindbergh's success and designed to help her acting career. In part, they were probably right. The publicity generated by her announcement was good exposure for her career. But it was more than that. Elder was out to prove that a woman was equal to a man. It was that simple. The ocean crossing that lay ahead of her was far from simple.

Ruth Elder was a very deliberate person. In plotting her routes, she made doubly sure to avoid the worst of the Atlantic storms. But that wasn't good enough. In her headlong approach to this goal, she ignored some basic advice - avoid the North Atlantic in winter. Sure, Lindbergh had succeeded, but perhaps he was lucky. Everyone before him had tried and failed. Elder's backers urged her to wait until spring but other women were preparing to attempt the flight also. She didn't want to lose out to one of them; she could taste the victory. She was going to be the first. (Fig. 2-3)

Elder chose a Stinson "Detroiter" airplane, proven for its ability in long distance flying. She called it the *American Girl*. "Looking back," she said, "perhaps my drive to succeed clouded my judgement. The weather was awful. My choice of copilot, George Halderman, was as deliberate as my choice of airplane. He was one of the best pilots of the day." Because of the rash of accidents that had occurred at Roosevelt Field, Long Island, plus the fact that Elder did not have a pilot's license, the owner of the field refused to let her take off. When Elder proved her ability as a pilot, the owner of the field backed off but only if her copilot, Halderman, would pilot the plane and she would act as copilot.

Elder remembers, "On October 11, 1927, in spite of bad weather, we took off. The *American Girl* carried 520 gallons of fuel, enough for 48 hours of flying time." Lindbergh had made the flight in 21 hours, 40 minutes, and Elder felt the *American Girl* would make it even if they ran into worse weather conditions. The press at first did not take Elder seriously. Elder, they were sure, was just an attractive and liberated woman looking for publicity. They kept a low profile until they realized, on October 13, the *American Girl* was overdue. Then they splashed the front pages with headlines voicing concern and wishes for her safe arrival.

The newspapers sold out as soon as they hit the streets. The *New York Times* reported, "Everybody in France is eager to see this audacious girl succeed in proving that she is not a weak woman. If she does succeed, that lovely American will have a triumph as great as Lindbergh's. The daring and self confidence of that American girl has imbued public opinion with the conviction that she will succeed. There will be no repetition of the pessimistic predictions that sought to discourage flights since the recent scenes of transatlantic disasters."

Elder almost made the dangerous crossing successfully. The *American Girl* flew for 28 hours through storms during most of the trip over the Atlantic. Elder and Halderman flew within 360 miles of the Azores. An oil leak forced them to land in the water. Elder anticipated the possibility of a water landing and had charted her course near the active shipping lanes. A Dutch oil tanker rescued them a short time after they ditched. They found a tumultuous welcome in Paris, and again in New York. But not everyone hailed her valiant and brave attempt as heroic. Katherine Davis, a sociologist, agreed with many of the male attitudes about flying and said so publicly. "There is no woman alive today equipped for such a flight. She should not have even attempted such foolishness."[12]

In a few short years, Amelia Earhart would prove Davis embarrassingly wrong. Elder continued flying and in 1929, she came in fifth in the First Women's Air Derby. Elder soon after retired from aviation and went on to become a successful Hollywood actress.

Fig. 2-3
Ruth Elder

[12] Judy Lomax. *Women of the Air*. Dodd, Mead, & Company, N.Y. 1987; p.66

Record Fever

Women had been flying for almost two decades and yet the record books did not reflect their achievements. Viola Gentry's attempt to break a man's endurance record in 1928 provided the impetus for a series of record challenges that followed.

Gentry, another believer in the "lucky number 13," had decided she would remain aloft for 13 hours, 13 minutes, and 13 seconds. On December 2, 1928, she took off from Curtiss Field, on Long Island. For the long winter flight she wore two flying outfits. That did not help much when bad weather rolled in, forcing her down after being in the air for eight hours, six minutes 37 seconds. Her flight did not exceed any men's record, but it was the first formal attempt by a woman to break a man's record. The prestigious Federation Aeronautique International (FAI) sanctioned and monitored the event. When the FAI recorded the attempt, the flood gates opened.

Within two weeks, Evelyn "Bobbi" Trout broke Gentry's "record" by staying aloft 12 hours, 11 minutes, and the flight went down in the books as the longest by a woman to date. Three weeks later, Elinor Smith broke Trout's record by staying airborne 13 hours, 16 minutes. Bobbi Trout recaptured the record a month later. A month after that, Louise Thaden set a new record of 22 hours, 3 minutes and a month later, 17 year-old Elinor Smith recaptured the honors with a 26 hour, 21 minute flight. Record fever was high in 1929 but the women had a long way to go if they were to break the men's record of 60 hours. [13]

Matilde Moisant recalled, "We were setting new records among women pilots but they were all unofficial. The records did give us some benchmark to look back on when any of us set out to establish a new record." Louise Thaden, for example, set the first official woman's altitude record of 20,260 feet on December 7, 1928, but it only stood until May 28, 1929, when Marvel Crosson reached 24,000 feet.

The records of these air pioneers are now recognized and recorded, but again, not before a struggle. At the FAI conference in Copenhagen Denmark, in 1929, there was a long debate about the need for a separate category and recognition of women's events. Women's achievements were no longer compared to men's and could be put in proper perspective. This gave the women time to catch up in skill, and gain access to airplanes that would allow them to compete with men if they chose. The establishment of a woman's category in aviation motivated more women to enter the field and in doing so helped women advance their skills.

At first women competed against each other in their own races, but by the mid-1930s they had honed their skills and were flying against men in events like the transcontinental Bendix Air Race. Louise Thaden and Blanche Noyes won the race in 1936, and Laura Ingells finished second.

---

[13]   Kathleen Brooks-Pazmany. *United States Women in Aviation, 1919-1929.* Smithsonian Studies In Air and Space, #5; p.30.

In 1938, Jacqueline Cochran captured first place in the same race. The men had met their equals and most of them did not like the feeling.

Ruth Law's name was synonymous with stunt flying. Phoebe Fairgrave Omlie achieved similar fame as a stunt flier for the movies for her piloting in "The Perils of Pauline." Elinor Smith, at age seventeen, earned international acclaim and a reprimand from the Department of Commerce for flying under all four of the East River Bridges in New York City. Smith, Viola Gentry, and Bobbie Trout outdid each other in setting new endurance records for women. Trout and Smith were the first civilian pilots to refuel in midair. In January 1929, they stayed in the air for 45 hours and 5 minutes. In January 1931, Trout and Edna May Cooper set another refueling record of 122 hours and 20 minutes. In August 1932, Louise Thaden and Frances Marsalis stayed aloft for more than eight days.

The number of women pilots increased during the barnstorming era, and more records were set and broken by women. Ruth Nichols believed that the purpose of a record was to break it and wished that, "More girls should get good ships and keep setting new marks. It has long been my theory," she said, "that if women could set records often duplicating the men's, the general public would have more confidence in aviation." [14]

Women's Air Derby

One of the last male bastions of the 1920s was competition flying. It fell as the decade came to a close. Perhaps the most singular event to occur in the two decades women had been flying that forced men to accept women as serious pilots was the Women's Air Derby, the opening attraction to the August 1929 Air Races, in Cleveland, Ohio. The race started in Santa Monica, California, and ended a week later, 2,800 miles away at Cleveland's Municipal Airport. The first prize was $2,500, and it attracted considerable interest. The public would later recognize the race as a grueling test of endurance, flying ability and courage.

To be eligible, women pilots had to have a license and at least 100 hours of solo time, 25 of which had to be cross-country flights of 40 miles or more. Of the 40 American women who could qualify, 18 took part. There were also two European women entered. The women were quickly and frivolously called "Petticoat Pilots," "Angels," and the "Flying Flappers." The Derby became known as the "Powder Puff Derby," and ironically the women later adopted that name for the event.

There was constant danger throughout the race and the pilots had to take along a gallon of drinking water, a three-day food supply, and a parachute. To navigate they used road maps, temperamental compasses, dead reckoning, and plain old seat-of-the-pants flying. Heat-haze, blowing sand, and driving rain added to their navigation problems. The terrain varied from parching deserts to the frigid peaks of the Rocky Mountains,

---

[14] "Flair For Flight." *Ninety-Nine News.* March-April, 1974, p.10

forested hills, and lush green valleys. They also had to contend with last minute route changes, sabotage, and death.

Tragedy marred the event. Ruth Nichols' close friend, Marvel Crosson, died after she bailed out too low for her parachute to open. Some of the newspapers used the incident to reinforce their chauvinistic views, and one stated sarcastically, in bold headlines. "Women Have Conclusively Proven That They Cannot fly."[15]

J.D. Halliburton, a Texas oilman, said, "Women have been dependent on men for guidance for so long that when they are put to their own resources, they are handicapped. Stop the race, for the sake of the women."[16]

Frank Copeland, the race manager, responded very bluntly to Halliburton. "We wish to thumb our collective noses at Mr. Halliburton," he said. "There will be no stopping this race!"

The women cheered the decision and continued the race. Fourteen exhausted but triumphant women proved Halliburton wrong when they crossed the finish line. Beside the one fatality, one contestant dropped out two days into the race when she came down with a high fever and symptoms of typhoid. Three others had accidents, and although uninjured, dropped out with aircraft damaged beyond repair. Ruth Nichols, who had been in the lead dropped out on the last morning of the race. On take-off, a strong crosswind forced her into a tractor, and somersaulted her airplane several times before it came to a stop. Nichols, uninjured, crawled out from the pile of junk that had once been her airplane.

Twenty-three year old Louise Thaden, the winner, had a very pragmatic view of life and of flying. "My success in the Derby was more important than life or death. We women were out to prove that flying is safe. I think the results proved that our purpose had been more than adequately fulfilled."[17] The women finished the race, but the problems the women encountered were universal rather than specific to gender. The point women were also trying to make for twenty years had finally been made. Flying was not exclusively a male domain.

There was also another benefit with far reaching implications that grew out of the Derby. The races provided women with another avenue for exposure and yielded a significant return. The women's Derby gave the industry another forum to develop the state-of-the-art and those who participated became in effect, the first women test pilots. Like their male counterparts they tried untested designs, new fuels, and instruments. Many of their findings eventually translated into commercial and military applications, with very positive results.

[15] Judy Lomax. *Women of the Air.* Dodd, Mead, & Company, N.Y. 1987; p.60

[16] Kathleen Brooks Pazmany. *United States Women in Aviation 1919-1929.* Smithsonian Studies in Air & Space #5; p.46.

[17] ibid, p.51

The Ninety-Nines

The Cleveland Air Races had drawn woman pilots from all over the country. It was the first time many of them had met face-to-face. When they did, a special relationship developed almost immediately. They discussed careers, traded hangar talk, and realized what was missing from their lives; unity. The more they talked, the more enthusiastic they became. An organization where they could share experiences, encourage other women to take up aviation careers, and maybe even become politically active appeared necessary and possible.

Clara (Studer) Trenckmann, not a pilot herself but a strong believer of women in aviation, set the ball in motion. Clara had been working at the Curtiss Flying Service and was the editor of the company's newsletter. On October 1, 1929, she asked Curtiss executives to allow four Curtiss demonstration pilots, Fay Gillis, Neva Paris, Frances Harrell, and Margery Brown, to draft a letter to all licensed women pilots inviting them to a meeting at Curtiss Field, in Valley Stream, Long Island. The letter set the meeting date for November 2, at 3 p.m.. Because of poor weather, only four women flew in and another 22 arrived by train. Clara provided an invaluable service. She handled the responses, arranged for the meeting place, and provided the food and transportation for the out-of-town pilots. She converted conversation into action.

They met in the hangar where conversation was difficult because occasionally someone would start an engine just outside the hangar doors. A mechanic repairing a nearby plane, hissing spray guns, and the smell of paint did not help matters either.

The women saw the seed of an organization developing that would provide a close relationship among the women pilots, and help them in aeronautical research. This organization would identify air racing events, help women with jobs in the aviation industry, and provide airlifts in times of emergency arising from fire, famine, flood, and war. The organization would also provide any other interest that would benefit women and be of general aviation nature. They emerged several hours later with the foundation of the Women Pilots Association.

The women decided that the club would serve as a social and professional organization, and would be open to licensed women pilots exclusively. The women were also concerned about the poor showing. They were disappointed until they discovered they did not have a complete list of names. Neva Paris and Amelia Earhart drafted a second letter with the invitation to meet at Opal Kunz's home in New York City, on December 14, 1929.

At the second meeting there were hours of discussion on a name for the group. A pilot named Jean Davis Hoyt suggested naming the club after the number of women at the meeting plus those who did not attend but wrote, telephoned, or telegraphed their interest in joining. The total

Fig. 2-4 Twenty-two of the original Ninety-Nines.

came to ninety-nine names, and so the Ninety-Nines became their official name (Fig. 2-4). [18]

Leadership problems arose immediately. These were strong-willed women who did not want to take second spot to anyone. They finally agreed to hold elections with Neva Paris coordinating them, but Paris died in an air accident enroute to an air race in Florida. The club remained unstructured while Louise Thaden served as secretary. In 1931, the group voted Amelia Earhart as the Ninety-Nines' first president.

18 ibid, p.52

The Ninety-Nines needed favorable publicity for their fledgling organization. In December, 1930, four women simulated an air raid on New York City, bombarding the city with thousands of leaflets. The leaflets appealed for funds to support the Salvation Army's unemployment relief, and advertised a charity supper-dance organized by the Ninety-Nines. The dinner was an overwhelming success.

In 1933, the Ninety-Nines flew a similar publicity fly-by for Roosevelt's National Reconstruction Act (NRA). The *New York Times* reported, "Fifteen feminine flying aces thrilled thousands yesterday with an NRA Air Pageant over Manhattan. Ten planes led by Elinor Smith, carried the fliers through maneuvers, and for a finale, dropped bouquets."

Betty Gilles, one of the pilots, was angry. "Not everyone was as impressed," she said. "I thought we were a disorganized formation of pilots over New York City. I was one of them. It was terrible. Never again!"[19]

A long-lasting achievement of the Ninety Nines was the successful completion of an aerial marking campaign across America. Phoebe Fairgrave Omlie suggested that air routes without markings were like highways without signs. As a member of the National Advisory Committee for Aeronautics in 1936, she was influential in getting people to listen to her suggestion. Omlie gained the support of one very influential person, Eleanor Roosevelt. Mrs. Roosevelt gently influenced the Bureau of Air Commerce to hire women pilots to find suitable buildings, and negotiate with local officials and owners. The selected buildings would be painted with the place name and the distance and direction to the nearest airfield. The bold orange letters would be large enough to be seen from 3,000 feet. The Ninety-Nines painted more than 16,000 roof-tops to guide pilots over the vast American landscape. When World War II started, the signs were painted out to prevent the enemy from navigating the skies over America. After the war, many of the roof-tops were repainted.

In the early days, the Ninety-Nines did everything they could to promote aviation and stay solvent. They cooked pancake breakfasts, had cookie sales, washed airplanes and set up displays and flew at local air shows.

The Ninety-Nines have become a formidable organization in furthering women's commercial interests in aviation. It serves as a network and inspiration for women to write articles on aviation and foster a sense of air-mindedness. The Ninety-Nines address civic clubs, schools, and governmental institutions, taking their knowledge and love of aviation to the community-at-large.

Today the Ninety-Nines number more than 7,000 members in nine countries. In 1979, when the Ninety-Nines celebrated their Golden Jubilee, Air Force General (and Senator) Barry Goldwater hailed, "...

---

[19] Judy Lomax. *Women of the Air*. Dodd, Mead, & Company, N.Y. 1987; p.62

the great accomplishments of women pilots throughout the last half century, particularly the lasting contribution they made to the development of aerospace pursuits." His words were proof of what Amelia Earhart had said fifty years earlier, "If enough of us keep trying, we'll get someplace."

Fig. 2-5 Ladybirds circa 1930. (L-R) Mrs. John Remey, Edith Descomb, Mrs. Freddie Lund, Helen Fitzgerald, Jean Lenore Stiles, Smaranda Braescu, Viola Gentry, Mrs. Robert Moore.

Chapter 3

The Golden Age of Aviation 1930-1941

Because of where I came from, and then where I went, I ended up understanding intimately, one very sustaining fact of life. I could never have so little that I hadn't had less.

Jacqueline Cochran

In the 1930s, the industry had succeeded in making the airplane a reliable means of transportation. The general public was still not convinced. It took the calculated introduction of women as pilots to prove to a skeptical public that airplanes were safe and easy to fly. Corporations hired women to race, tour, make special promotional flights, fly executives, and handle airplane sales to the private sector. Some have said that women aviators in the 1930s were manipulated by the corporations. There is no doubt that aircraft manufacturers hired women as pilots to show that if a woman could fly a plane anyone could fly. Most people agree that women were aware of this and most did not feel used. It was practically the only way they were able to fly. They allowed the exploitation to gain access to airplanes and they made the best of a bad situation.

Other women were beginning to emerge from the cocoon of Victorian ideas. A few were even beginning to see that it was okay to make career "mistakes," change their minds and allow and accept change as a God-given right. They had not been accepted or allowed the right to choose aviation as a career but aviation in the 1930s is where many emerged from their cocoon, with working wings.

In 1930, the role of American women in aviation was still marginal. Before the decade was over, however, women would make the first significant inroads into aviation. In 1932, there were 472 licensed women pilots in the United States. Of these about 50 held an Air Transport Pilot rating, the highest. Compared to the entire male pilot listing, it is less than three percent (17,226 in October 1931). If you take into consideration only the 50 Air Transport Pilot ratings, because none of the others at the time could be counted as offering serious commercial possibilities, the number of potential candidates among women for available jobs was very small.[1]

In the decade between 1930 and 1940, the "Golden Age of Aviation," major advances took place in aeronautics. Although the Golden Age officially kicked off with Lindbergh's successful transatlantic flight in 1927, recognition of women in aviation began to occur in the 1930s. The dangers of Barnstorming created federal laws restricting barnstorm pilots to minimum altitudes for demonstrations. The laws did not recog-

---

[1] Amelia Earhart, *The Fun of It*. New York, Harcourt Brace; 1932; p.146

nize that a large number of women pilots were barnstormers and largely forced out of business.

In spite of this, the 1930s marked the first positive change in the outlook for women in aviation. Amelia Earhart defined why women were flying. It was a dual message; flying is safe, and women make good pilots. The two ideas went hand in hand.[2]

By the early 1930s, the prejudice against women pilots was still rampant. Charles Lindbergh's wife, Anne Morrow, was a pilot herself and a strong proponent of women in aviation. Her husband still had his own narrow ideas. His position as a major figure in aviation gave his ideas and public statements credibility, and reinforced the popular notion that women did not belong in aviation. When he saw women flying in the Soviet Union's Air Force in 1938, he recorded in his diary, "I do not see how it can work very well. After all, there is a God-made difference between men and women that even the Soviet Union can't eradicate."[3]

Other influential persons in aviation were aware of women's efforts and accomplishments and could have helped to expand the roles of women in aviation. They were surprisingly restrictive in their views. Eddie Rickenbacker, World War I Ace and head of Eastern Air Transport, took the executives of Boeing to task in 1930 for hiring the first female airline flight attendants. He argued that flying was a man's occupation and should stay that way. Ironically, Ellen Church, the first flight attendant, was a pilot and seeking employment as such when Boeing hired her to serve food and look after passengers.

Lillian Gatlin

Women had been trying for over a decade to overcome this malignant attitude. By the third decade of aviation, women were taking to the air in increasing numbers, some first as passengers, and others as pilots. They were finally following Lillian Gatlin's lead. Gatlin was the first woman to make a transcontinental flight but she did so as a passenger. Gatlin, the president of the National Association of Gold Star Mothers,[4] made the first flight in a mailplane since there was no commercial service in October, 1922. Gatlin had hoped the flight would stimulate interest in having a day set aside as a memorial day to fliers killed in the war. She did not succeed, but later, November 11th became the day dedicated to honor all who died in World War I.

When Gatlin was questioned about her trip, she said, "I had a good deal of rest. Flying is the ideal method of traveling, no cinders, no invitation to buy any products advertised on sign boards extending from

---

[2] Claudia Oaks, *United States Women in Aviation 1930-1939*. Washington. Smithsonian Studies in Air and Space #6, p.4

[3] Valerie Moolman, *Women Aloft*. Va. Time Life Books, 1981; p.7

[4] An organization formed by mothers who lost a son in the War.

coast to coast, nothing to disturb the easy sailing through the atmosphere." All this enthusiasm from a woman who had just spent 27 hours, 11 minutes in an open cockpit mailplane that traveled 2,690 miles.[5]

Gatlin's flight had made little difference as women flew into the third decade of aviation. There were also well-meaning supporters of women pilots in the male community. A chief pilot for a large flying service had no doubt women made good pilots. In his own, but typical way, he explained, "They are easier to teach, and learn quicker than men. Women usually think about flying for a long time before they start taking instruction. They leave the instruction to you. When you tell them their mistakes, they pay more attention, and often correct them faster."[6]

The first attempts women made to fly across the oceans were also made as passengers, not pilots. Even today the oceans are terribly unforgiving of pilot errors, and women as passengers in the 1930s faced the same dangers and took the same risks as the men who flew the machines.

Louise Thaden and Frances Marsalis together had set an eight-day endurance record in 1932. "I see this question of a woman's ability to fly developing into the battle of the sexes. Women can never hope to compete with men," Thaden said, "in the actual flying of airplanes. Not that a woman can't handle a plane as well as a man. She can, and many of them do the job a lot better. But the public doesn't have the confidence in women fliers. That is, not enough confidence to ride with us to any great extent."

She continued, "This attitude on the part of John Public, and he'll never get over it, means that all women are forever barred from careers as transport pilots on regular passenger lines. Promotion and advertising, that's the field for women." Louise was disturbingly accurate in pinpointing the problem, at least for the next four decades.

The transcontinental Bendix Air Race, a landmark race and the most popular flying event of the 1930s, which began in 1931, was closed to women on an equal basis until 1935. (Amelia Earhart was allowed to enter in 1932 but all women were banned in 1934 because 25-year old Florence Klingensmith, had been killed flying a *Gee Bee* Racer, in a Chicago air meet in 1933.) In 1935, the races were limited to separate all-women events that were restricted to stock, commercially licensed aircraft with air speeds less than 150 mph. Race officials assumed women would not compete successfully with men in the grueling race, so they set aside a special $2,500 prize for the first woman to finish the race. It sounded like a consolation prize to the women. In 1936, Thaden, along with Blanche Noyes, flew to first place in the Bendix Air Race, in 14 hours, 55 minutes. That year, Thaden also won the Harmon Trophy as, "The World's Outstanding Flier." Since Louise Thaden and Blanche

---

[5] *New York Times,* October 9, 1922.

[6] Claudia Oaks, *United States Women in Aviation 1930-1939.* Washington. Smithsonian Studies in Air and Space #6, p.1

Noyes finished first, they walked away with the $4,500 and the $2,500 prize. To the astonishment of the officials, Laura Ingells finished second 45 minutes later. Amelia Earhart and Helen Richey, flying together, came in fifth. Two years later, Jacqueline Cochran won the race in the record time of 8 hrs. 10 min.

Anne Morrow Lindbergh

One woman in aviation who managed to stay out of her husband's enormous shadow and establish her own identity, was known as much for her writing as her flying; Anne Morrow Lindbergh. Today her husband Charles is still widely remembered, however, few people are aware of Anne's contributions to some of Lindbergh's flights. After she married America's most famous aviator in 1929, she lost little time in adopting her husband's aviation goals. Within three months she had soloed, and in 1930 she became the first woman in the U.S. to earn a glider pilot's license. She was her husband's copilot, navigator, and seven months pregnant in April 1930, when Charles Lindbergh set a transcontinental speed record. Although the route was surrounded by storms and she was painfully sick most of the trip she continued to do her part. In 1931, she received her private pilot's license. She was probably motivated toward this goal by the constant publicity and crowds. Flying with her husband was practically the only way they could have time alone together. As a true working partner, Anne often referred to herself as, "Charles' faithful page." Her husband said in turn, "No woman exists or has existed who is her equal."

Anne Lindbergh's most famous flight was the one shared with her husband, where they showed the feasibility of using the "Great Circle

Fig. 3-1 Anne Morrow Lindbergh

Route" straight up Canada, over the North Pole, to China. On the flight that she later described in the rich narrative, *North to the Orient,* and *Listen; The Wind,* she served as navigator, radio operator, and copilot. For her part in the study of international airline routes she received the United States Flag Association *Cross of Honor,* and in 1934 became the first woman to receive the National Geographic Society's *Hubbard Medal.* (Fig. 3-1)

In the 1950s, Anne wrote a special book. Gift From the Sea, was a group of essays on love, marriage, youth and old age, all inspired by sea shells. It spoke elegantly to women about a need to escape from the routines of life and seek self-fulfillment. In it she related what she had learned during her lifetime in her search for independence - that basically everyone is alone. Aviation fascinated Anne Lindbergh. "I was conscious of the fundamental magic of flying. It is a miracle that has nothing to do with any of its practical purposes of speed, accessibility, and convenience - and will not change as they change."

"For not only is life put into new patterns from the air, but it is somehow arrested, frozen into form . . . a glaze is put over life. There is no flow, no crack in the surface, a still reservoir, no ripple on its face."[8]

Amelia Earhart

There were a dozen giants among the women pilots in the 1930s who showed skills and endurance equal to and better than many men. The most popular and the most controversial figure was Amelia Earhart. Amelia Earhart is probably the best known pilot in the history of aviation, and no history of women in aviation would be complete without her. Earhart's career in aviation lasted only ten years but in that short period of self-exploration she created a legend to which her death added an unsolved and mysterious element. At twenty she had worked as a nurse in a Toronto, Canada hospital. There she developed a strong pacifist attitude. Earhart left nursing after the war, saying the only positive outcome of the war was the "inevitability of flying." By the time Earhart started flying at the age of twenty-three, she had developed strong convictions on the independence of women, an aversion to alcohol, and the belief that a woman's place was not in the home. (Fig. 3-2)

After the war Earhart enrolled as a medical student in Columbia University, but soon decided becoming a doctor wasn't for her. At an air show in Long Beach, California, she persuaded her father to give her $10 for an airplane ride. When she landed, Earhart knew what to do with her life.

Earhart began her lessons with Netta Snook, and after two hours of instruction, went out and bought a $2,000 Kinner "Sportsplane." Some

[7] ibid

[8] Anne Morrow Lindbergh, *North To the Orient.* New York. Harcourt, Brace, & World Inc. 1935; p.37

people say she took up flying to escape the tensions created at home by an alcoholic father, and the pressures of a persistent suitor, Sam Chapman, a boarder in her mother's house. Those who made those statements have not looked deeply into her personality. Earhart flew because she considered herself no less equal than a man and flying was one way to prove this point. The record speaks of the success she had in proving her point.

Although the slow-moving, dangerous, hydrogen-filled German dirigibles had been crossing the Atlantic, Earhart knew the Atlantic was still the major frontier for aviation. Until conquered by the airplane, there was no real international air service. Crossing the Atlantic was dangerous and deadly. There had been only six successful crossings, the first by Alcock and Whitten-Brown, together in 1919.

The year was 1927, and Gene Tunney was the heavyweight champ. F. Scott Fitzgerald described the opening of a new age in America. "Something bright and alien flashed across the sky. A young Minnesotan who seemed to have nothing to do with his generation did a heroic thing..." Charles Lindbergh had made the first solo crossing of the Atlantic in

Fig. 3-2 Amelia Earhart

May. The same year Lindbergh conquered the Atlantic, nineteen others had died in unsuccessful attempts. Three women were among the fatalities, each as a passenger of a male pilot.

The following year, in April, Earhart's life took a dramatic turn. The crew of the *Friendship,* a Fokker trimotor, invited her to join them as a passenger in an attempt to fly across the Atlantic. The flight was sponsored by a wealthy Pittsburgh heiress named Amy Phipps Guest. Mrs. Guest's ultimate purpose was to have a woman fly across the Atlantic by airplane. Until that fateful day, Earhart's flying career had been on again, off again, with a second attempt at medical school, a short stint at teaching, and a job as a social worker in an immigrant center in Boston. Earhart accepted the challenge, and the gold and flaming-red-painted *Friendship* took off on June 28, 1928. Earhart spent most of the flight wedged between two gas tanks, keeping a log and notebook of personal observations. She watched the pilot as he maneuvered through fog, rain and some of the worst storms in the north Atlantic. She was apprenticing his every move. During the long, cold, and dangerous trip they lost radio contact, flew blind just 300 feet over the waves, and missed the target, Ireland, but landed in Wales with about a gallon of fuel remaining. Earhart had apprenticed with the best. The *Friendship's* pilot, Wilmer Stultz, received $20,000; the navigator, Louis Gordon, $5,000; and Earhart nothing. She had gone along, for "the fun of it."[9]

Earhart had become the first woman to fly across the Atlantic in an airplane. While there was praise from many circles, there was criticism in the European press that attempted to dilute Earhart's accomplishment. They congratulated her on escaping the fate of three of her peers who had made fatal attempts at the same goal. Some grudgingly lauded Earhart as an international heroine based on luck and the skill of a male pilot. One paper said her presence added no more to the achievement than if the passenger had been a sheep.[10]

The *New York Times* finally conceded a point for women when they gave their opinion on the importance of her presence. "What transatlantic aviation needs if it is ever to reach the commercial stage. . . (is) something in the nature of a series of carefully engineered experiments. Because of the circumstances under which she flies, Miss Earhart's must be regarded as the first step in an engineering investigation out of which will emerge the practical transatlantic passenger carrying flier of the future."[11]

In America, the press called Earhart "Lady Lindy," and she became the unwilling captive of the public domain. George Putnam, a publisher, (and later her husband) had orchestrated her publicity with the idea of writing a follow-up best seller to Lindbergh's personal account of his

9 Later the title of her book about the adventure.

10 Charles Paul May. *Women in Aeronautics*. New York, Thomas Nelson & Sons Inc. 1962; p.73

11 *New York Times*. June,30 1928.

transatlantic flight. Putnam began manipulating Earhart's public appearances and receptions, deciding what invitations she would accept, and whom she would meet. He organized an exhaustive schedule of twenty-seven appearances in one month. The public's curiosity was piqued. They wanted to hear more about the dangerous crossing and see the remarkable physical resemblance to Lindbergh for themselves.

The instant celebrity status catapulted her into long-lasting headlines. Earhart began to see the advantage in the situation, and became an articulate spokesperson for women and aviation. Her deep commitment to women's potential in aviation now had a forum. Her popularity grew quickly, and Earhart's public forum grew larger. In 1928, she became an associate editor of *Cosmopolitan*, writing articles keyed on aviation themes for women like, "Shall You Let your Daughter Fly?" The answer was an unequivocal, Yes!

Even strong people occasionally have self-doubt, and Amelia was no different. She felt her reputation was undeserved although she was one of the top woman fliers in the United States. She was the first to make a transcontinental flight in an autogiro, an early helicopter-like vehicle. There was still the challenge of flying the Atlantic alone, something Earhart felt she had to do. On Friday, May 20, 1932 she took off alone in her Lockheed *Vega,* from Harbor Grace, Newfoundland, for Ireland, to personally earn her wings and reputation. The flight was without major coverage, but not without excitement and danger. Earhart had estimated the flight would cover 1,860 miles, but because of weather and instrument problems she flew 2,065 miles about 500 miles short of Lindbergh's record setting flight. There were dangerous electrical storms. A weld on an exhaust stack burned through, causing severe vibrations. Her barograph measured a 3,000 foot vertical dive from a spin induced by iced wings. Her altimeter malfunctioned when she was flying blind, but in spite of the hazards, she landed safely, and her reputation and immortality were assured.

In 1935, Earhart flew the same *Vega* from Honolulu to Oakland, California in 18 hours, 16 minutes, and became the first person to solo between those two points. All that remained for Earhart or any woman, it seemed, was to fly a true around-the-world flight. Up to this point, attempts at round-the-world flights always occurred north of the equator, where there was a predominance of land, and the water expanse was not as great.

As a visiting faculty member at Purdue University, in 1937, she had a twin-engine Lockheed *Electra* placed at her disposal. When Earhart accepted the plane she said, "I have just one more good long distance flight left in my system." That statement was an omen of the future.

Because it represented the most risk and the greatest distance, no one had flown around the world at the Equator. Earhart planned to fly an east-to-west flight across the Pacific to Honolulu, and continue as close to the Equator as possible, covering some 27,000 miles. With Paul Mantz as her copilot and Fred Noonan and Harry Manning as navigators, she made Honolulu without any problems. On takeoff from Luke Field, Hawaii, with only Manning on board, a tire blew and the plane ground-looped. The plane needed extensive repair and went by boat to the mainland. After studying the global weather conditions Noonan decided that a second attempt with a June 2nd departure would favor their trip.

The trip went well for forty days, and 22,000 miles, until the longest and most dangerous leg, a 2,556 mile hop from Lae, New Guinea, to Howland Island, which was a one-mile by two-mile strip of sand near the Equator. To ensure her safety, the U.S. Navy had ships along the route with the latest radio direction finding equipment. There were also other surface ships stationed along the route for an emergency. Earhart sent several messages enroute to Howland Island, but the plane never arrived. The last message received from her read, "We are on the line of position 157-337... We are running north and south." This message did not last long enough for anyone to pinpoint her position, and by itself was meaningless. Officials knew she had missed the small island target, was low on fuel, and unable to receive messages coming from the Coast Guard Cutter *Itasca* believed to be in the vicinity. They pictured a calm, but worried Earhart and Noonan frantically trying to locate a land mass or radio signal.

President Roosevelt ordered a massive 260,000 square mile air-sea search costing more than $4 million, but it failed to turn up the plane or the pilots. On July 18, 1937, the U.S. Navy declared Earhart and Noonan lost at sea.

The subsequent investigation found that carelessness had led to the tragedy. Earhart had left the emergency flares in the hangar and had removed some radio equipment to reduce the plane's weight. In doing so, she had reduced the range of the equipment. The aircraft also lacked an emergency portable radio, and the official report said, "It is very apparent that the weak link in the combination was the crew's lack of expert knowledge in radio." No one, it appears, told Earhart about other radio frequencies available, or about the high frequency radio direction finding equipment on Howland Island.

Earhart's mysterious disappearance still generates speculation. Many rumors persist, the strongest being she was a spy for the United States and executed by the Japanese because she had discovered the military build-up in the Pacific. While that rumor persists, it is unlikely, since Earhart was an ardent pacifist and not trained in spy activities. As for spy equipment, she carried only a tourist-type camera on board. Perhaps Earhart was a spy. It made more interesting reading than to say she and Noonan simply ran out of gas and were lost at sea. (The goal Amelia was reaching for was finally reached in March 1964, when Geraldine "Jerrie" Mock flew her single-engine Cessna 180 around the world.)

Louise Thaden did not think Earhart's death was a bad way to go. "If your time has come," she said, "it is a glorious way to pass over. The smell of burning oil, the feel of strength and power beneath your hands, so quick has been the transition between life and death there still must linger in your mind's eye the everlasting beauty and joy of flight. Women pilots were blazing a new trail. Each pioneering effort must bow to death. There has never been, nor will there ever be, progress without sacrifice of human life."[12]

Thaden said, "Like the rest of us, Amelia had ambitions. Unlike most of us she had a definite notion of each progressive step toward the set of goals. Frustration, hundreds of smaller obstacles, but probably most of all loneliness, could not deter her from ascending the pinnacle of predetermined achievement."

Thaden continued, "It may seem incongruous, yet Amelia Earhart's personal ambitions were secondary to the insatiable desire to get women into the air, and once in the air to have recognition she felt they deserved, accorded them."[13]

Earhart's male critics found her overconfident and arrogant; they accused her of using her husband's influence in high places to get her way. They also did not consider her a good pilot. On that account her pilot skills need no defense, her record speaks for itself. She was the first person to solo from Hawaii to California; from California to Mexico; and from Mexico to Newark, New Jersey. As a writer and public speaker she made millions of friends for aviation, and influenced women then and now to grow wings.

To her admirers then and now, Earhart is a symbol of the emancipated woman in the forefront of women's rights. She was an equal to men on the ground and in the air, and did not need her husband's influence. Her existence and involvement in aviation extended to many women opportunities and new possibilities.She was a strong role model and her writing added reinforcement to the fact that a woman could do anything she set her mind to, even fly - like the men. Amelia Earhart had the courage to change even after an emotional or financial investment, and the drive and self-assurance to dare anything.

Jacqueline Cochran summed it up a few years later, "I think flying takes courage ... We must remember that courage, to be important, must endure. Anybody can take anything for a little while. But when you're blazing new trails you have to be able to endure."[14]

Jacqueline Cochran

In the summer of 1912, the year Harriet Quimby died in the Boston Air Meet, a baby was born in the Florida Panhandle. This baby would grow up as if she were the reincarnated image of Quimby's energy, and fulfill many of Harriet Quimby's dreams.

Jacqueline Cochran was an authentic American hero. She was the first woman to fly a bomber across the Atlantic Ocean to England, the first woman to fly a jet across the Atlantic, and the first woman to break the

[12]
Louise Thaden. *High Wide and Frightened.* New York, Stackpole & Sons. 1938; p.76

[13]
Claudia Oaks. *United States Women in Aviation 1930-1939.* Washington. Smithsonian Studies in Air and Space #6 p.27

[14]
ibid

sound barrier. There were few female test pilots and none who gained the fame of Jacqueline Cochran. She was a test pilot who constantly pushed the edge of the envelope, and routinely risked her life in unproven aircraft. She died in 1980, and today few Americans outside aviation know her name. Others are still trying to forget her. (Fig. 3-3)

Her career spanned more than 40 years, and when she died, Jackie Cochran held more speed, altitude, and distance records than any pilot, male or female. Cochran to some was a contradiction in terms. She was self-confident, demanding, at times arrogant, and also feminine, and loving. She was also called, "The greatest woman pilot in aviation history." Cochran went without shoes until she was eight years old. Her bed was a mat on the floor, and her diet at best was mullet, sowbelly, and blackeyed peas. During hard times, she foraged in the woods and fields for pine cone nuts, fished for perch or crab, or captured "stray" chickens. "Because of where I came from," she said, "and then where I went, I ended up understanding intimately, one very sustaining fact of life. I could never have so little that I hadn't had less."

Cochran never knew her real parents, and picked her name from a Pensacola phone book. She lived with a poor itinerant family in a shack on stilts, near a swamp, in Florida's Panhandle. Her adopted "father" and two "brothers" worked in a saw mill and received wood chips in payment for their labor. The chips were redeemable only at the company's store, and there were never enough chips on payday to meet the week's expenses. Cochran's family was eternally in debt to the "company store."

Fig. 3-3 Jacqueline Cochran

The company's doctor charged extra to deliver babies so Cochran became a midwife before she had reached puberty. "I delivered babies" she said, "before I even knew that the stork was a bird." Cochran's incredible determination and strong personality originated in her dirt-poor beginnings.

"Women pilots of the Thirties were a very special breed. Most were born rich. I hadn't been born to such surroundings, but if I had to push my way in and pull my way to the top I'd do it," she said.[15]

Cochran put on her first pair of shoes when she went to work in the sawmill, and she wore nothing more than a flour sack dress. She aged tough and prematurely, and at age ten was earning six cents an hour for a 12-hour shift, supervising 15 other children in a Georgia cotton mill. During an employees' strike, she found work as live-in help for a beauty shop owner.

By the age of 15, she had learned the beautician's trade and saved enough money to buy a car and attend three years of nursing school. Formal schooling didn't come easy to Cochran. "I learned my ABCs by studying the railroad boxcars," she said, "and began to figure out the words. This was my first adventure into the literary world, and it frightened me."[16]

After three years of study at a local hospital, Cochran accepted a job with a country doctor in Bonifay, Florida. Nursing barely provided her with enough to live on, and she found it depressing. Reality struck Cochran after delivering a baby by a corncob and oil "mojo" lamp in a shack. The family did not even have a piece of cloth to wrap around the infant. Cochran knew then she could not relieve the suffering or brighten the lives of her fellow Floridians. She returned to her first vocation, and it soon earned her enough to buy a half interest in a Pensacola beauty shop. After helping build the business, she sold her interest and moved to New York City. She talked her way into a job at the Saks Fifth Avenue beauty shop of Antoine's, and was soon in charge of his Miami Beach store. One day in 1932, Cochran was dining in a Miami restaurant and seated alongside a Wall Street millionaire-financier named Floyd Odlum. She started talking about her dream to open her own cosmetics business and escape the confines of the beauty salon. Odlum was sympathetic. He was the son of a Methodist parson and had done his share of manual labor, working his way through college and law school. Odlum had set up his own business and by 1928, when he was thirty-six years old, he had total assets of over $6 million.

It was Odlum who planted the seed in Jackie's head about flying. He recognized that she was a good salesperson and would do well in any business. Odlum pointed out that the Depression had made every business very competitive and she would need wings to keep up with her competition. After his inadvertent piece of business advice, he became an ardent supporter of Jackie and soon her suitor.

Cochran took his off-handed comment seriously. Flying would help beat the other sales people to the punch. In the Summer of 1932, while crowds cheered Amelia Earhart' solo flight across the Atlantic, Cochran went to Roosevelt Field, Long Island, to take flying lessons. With the

---

[15] ibid

[16] Gene Burnett, "A Florida Native Born To Fly." *Florida Living.* September, 1977; p.88

same determination that enabled her to survive the sawmill circuit in the Florida Panhandle, she soloed in three days and passed her pilot license exam in two weeks. Because of the difficulty Cochran had with reading and writing, she never took a written test. She took every exam orally, a task some say is three times harder. From that point her life took a turn, literally in an upward direction.

She said later, "When I paid for my first lesson, the beauty operator ceased to exist and an aviator was born." A few months later she left Antoine's and headed for San Diego. There she persuaded a Navy flight instructor to give her the equivalent of the entire U.S. Navy flight training course. Within two years, she had established Jacqueline Cochran Cosmetics, with a salon in Chicago, and a cosmetics laboratory in New Jersey. The cosmetics business provided enough income to support her first love, flying.

On the advice of a friend in Pensacola, she went out and bought a used *Travelair* airplane for $1,200 and flew to Montreal for her first air meet. Later she sold the *Travelair* and bought a new *Waco* for $3,500, and sailed to England for the famous, and grueling MacRobertson London-to-Melbourne air race. Unfortunately the plane she intended to use for the race crashed on its delivery flight. "I was the only American woman in the race and I wasn't going to let the fact that I didn't have a plane keep me out of it," Cochran said in her own unique style. Unfortunately she borrowed one of the Granville brothers' *Gee Bee* Racers. The often deadly *Gee Bee* true to its reputation, developed one of its known quirks overheating - and the supercharger failed. Jacqueline Cochran dropped out of the race in Romania when the *Gee Bee* "Q.E.D." (Latin for "Quod Erat Demonstrandum" - Quite Easily Done) was not so easily flown. Of course we know now that the *Gee Bee* violated some basic rules of aerodynamics, and its stall characteristics were unpredictable. Cochran said, "They were killers. Few pilots flew a *Gee Bee* and lived to talk about it. Jimmy Doolittle was one, and I was another."

In the early 1930s, flying was still experimental and dangerous. Only a few thousand men - and roughly 400 women - had pilot's licenses. "The reason I went for record breaking and long-distance flying," said Cochran, "was simply then as now, the jobs as test pilots and airline pilots went to men, not women. The chances were that if a woman was selected for this training, before she had returned a profit on the heavy investment in such training, she would have converted herself into a wife and mother and stopped working.[17]

[17] Jacqueline Cochran; Maryann Bucknum Brinley: *Jackie Cochran The Autobiography of the Greatest Woman Pilot in Aviation History.* New York, Bantam Books. 1987;

"I had a choice between piloting light machines which were boring and cost money, or getting hold of fast, up-to-date aircraft in which I could try to break records. I might risk my neck but I could probably earn a living."[18]

Jackie Cochran was a relentless competitor in the air and just as impassioned about her womanhood. Cochran felt women were special, with distinct abilities that she felt made them excellent pilots. "Women have the advantage of keen sensitivity and intuition," she said. "They often show more patience than men. They are accurate in calculations and have a fine sense of detail which nicely equips them for difficult flying."[19]

Cochran took a "Staggerwing" a fancy name for an open cockpit, fabric-covered biplane with no heat, pressurization, or oxygen to an unheard of altitude of 33,000 feet, something no man had even attempted at the time. She ruptured a sinus blood vessel, got frostbite, almost froze to death, and became so disoriented from lack of oxygen it took her more than an hour of flying at a lower altitude to regain enough equilibrium to land the plane safely. Her pioneering efforts led to mandatory pressurization of cabins, and oxygen masks on future high altitude aircraft. Some pilots, like Wiley Post and Howard Hughes, recognized her singular efforts and gave her advice, comfort, and valuable instruction.

In 1935, she persuaded Bendix Race officials to reopen the prestigious race to women. She entered the race, but dropped out after losing her radio antenna a runway fence on takeoff. Amelia Earhart, the only other woman in the race, came in fifth and won $500. On May 10, 1936, Cochran's cosmetics business was flourishing, and her strongest supporter and admirer, Floyd Odlum, not a pilot himself, became her husband. His interests in aviation ranged from investments to sharing Jackie's victories.

In 1938, she competed in a field of eleven men, and finally won the Bendix Race she had fought so hard to have opened to women. Even that victory wasn't easy, and her grit was the winning factor. The manufacturer had left a wad of paper in the fuel line and she flew most of the race with the plane tilted so the wing tank could feed the engine.

Amelia Earhart, a close friend of Jackie often stayed at the Odlum Ranch, outside Palm Springs, California. She visited the ranch to rest and prepare for her last and fatal, round-the-world flight in 1937. Cochran played an interesting role in the search for Earhart. Cochran, who claimed to have ESP, later revealed she had a strong premonition about Earhart's upcoming flight. To quell her fears, Cochran bought Amelia a bright colored kite, a package of fish hooks, and a Swiss Army knife, in case she ditched at sea. In exchange, Earhart gave Cochran a cherished

18
    Gene Burnett, "A Florida Native Born To Fly." *Florida Living.* September, 1979; p.88

19
    Sally Van Wagenen Keil. *Those Wonderful Woman in Their Flying Machines.* New York Rawson Wade Publishers. 1979 p.47

possession, a small, silk American flag. Unknowingly, Earhart was passing the torch to her friend.

Cochran tried to persuade her friend to postpone her flight. Her premonitions in the past had always been disturbingly accurate. Her husband studied the psychic phenomena that Jackie reluctantly claimed she had and concluded his wife did have an extrasensory perception. He claimed the two of them communicated over distances, and even in their sleep. More than once, Cochran was able to describe accurately what Floyd was doing and wearing when he was miles away.[20]

When Amelia Earhart disappeared, her husband, George Putnam, persuaded Cochran to use her powers to help during the search. Cochran gave an account of what was happening to Earhart and Noonan. She said the aircraft was still floating and both pilots were alive, although Noonan was unconscious with a fractured skull. She described in precise detail the position of an American Coast Guard Cutter, the *Itasca*, and said it was near Earhart's plane. Cochran, who was a devout Catholic and firm believer in an after-life, went to church and lit a candle for her friend's soul. The accuracy of her descriptions remain untested. Cochran did claim she had an ESP for "living things," and the experience left her so emotionally drained that she never used the powers again. Putnam, convinced his wife died at sea, remarried eighteen months later.

By 1941, Cochran had gained preeminence in aviation, having by that time earned seventeen aviation records. General Henry "Hap" Arnold, Chief of the Army Air Force, suggested she recruit some women pilots and get over to England and fly for the Air Transport Auxiliary of the British Ferry Command which desperately needed qualified pilots to ferry planes in England.

On June 17, 1941, Cochran became the first woman to pilot a bomber (a Lockheed Hudson) across the Atlantic ocean, not an easy job since dozens of pilots had been lost in the Atlantic because of weather. But male bastions crumble slowly. Cochran was not allowed to take off or land the plane; a male co-pilot did that job. There were two reasons for this. First, General Arnold did not think, "A slip of a young girl could handle the controls of a bomber." [21](He later changed his mind and acknowledged publicly that women could fly as well as men. The other reason was the civilian male ferry pilots threatened to strike if she took off. In 1943, Nancy Love and Betty Gilles were scheduled to ferry a B-17 to Scotland. When they reached Goose Bay, Labrador, the flight was canceled by Arnold. Arnold had ordered that no woman fly transatlantic planes until he had time to study the matter. He never approved such flights. It would have been political suicide for Arnold. It was unthinkable to allow women "so close" to a combat zone, at least in the air, and, Arnold had seen the backlash from the British public when a woman Air

[20] Judy Lomax, *Women of the Air*. New York, Dodd, Mead, & Company. 1987;

[21] Michael Beebe. "Years of No Recognition Still Prone to Bee Sting for WASP." *Ledger-Star*, March 9, 1977.

Transport Auxiliary pilot had been shot down. He was not ready to deal with an angry American public.

Cochran succeeded in almost every undertaking. In one area, - politics, she was unsuccessful. Her friends urged Cochran to enter politics but she turned down two offers to run as the Democratic Congressional candidate. Cochran then tried unsuccessfully as a Republican candidate. Jackie Cochran had devoted her life to aviation and did not have the experience or glib tongue required of a politician. She quietly backed out of politics but not before leading a campaign to nominate General Eisenhower the Republican candidate for president. She coined the phrase, "I Like Ike," at a meeting in her New York apartment. Eisenhower made sure Jackie was on the platform when he gave his acceptance speech, and as president, he and Mamie stayed as guests at the Odlum ranch.

Cochran lobbied behind the scenes for the creation of an Air Force independent of the Army, and played a decisive role in that happening in 1947. Two years later, Cochran was flying the narrow and dangerous air corridors of Berlin with the Air Force, during the airlift.

In 1947, Cochran also met Captain Charles (Chuck) Yeager, the first American to break the sound barrier. It was the beginning of a long friendship between Yeager, Cochran, and their spouses.

By the early 1950s, jet fighters were on the scene and Cochran's ambition was to fly one. Floyd Odlum owned the company that built the North American Sabre jet, but even with that much influence it took two years to obtain permission for a civilian to fly the jet. Cochran's appointment as a consultant to Air Canada was the influence she needed to get into the cockpit of a Canadian Air Force F-86 Sabre jet.

Chuck Yeager prepared Jackie for the flight, and in May 1953, Cochran became the first woman to break the sound barrier. The one million dollars per flight hour insurance coverage reflected the danger of the flight. In six hours of flying, Cochran made 13 flights, and broke the sound barrier three times. In her first supersonic dive from 47,000 feet, she did not break the record. Undaunted, Cochran refueled and went up again, this time breaking the record. Then to satisfy the news media and *Life Magazine*, she and Yeager made a simultaneous flight. In a nearly vertical dive from 50,000 feet, they rocketed downward wingtip-to-wingtip at full throttle. Cochran later said it was a spiritual and emotional experience that left her speechless.

In her fifties, Cochran was still active in aviation and still breaking records. Throughout her life she had a variety of illnesses and almost a dozen major operations. She had seven on her stomach, three on her eyes, and one on her sinuses, but none of them slowed her down. In her sixties she flew a Lockheed Jetstar from Houston, Texas, to Hanover, England, and became the first woman to fly a jet across the Atlantic.

There were few honors in aviation or in business that she did not receive. In 1945, she received the Distinguished Service Medal. She also received the Distinguished Flying Cross three times, an honor normally awarded only to Armed Forces personnel. Cochran received the gold wings of the Federation Aeronautique International, and the International Harmon Trophy fourteen times. She rose to the rank of Colonel in the Air Force Reserve, received honorary wings from a half dozen air forces, the French Legion of Honor, several honorary degrees, and the "Woman of the Year" award several times.

In her sixties, a pacemaker forced her to give up flying. After her husband died at the age of 80, in 1977, her health deteriorated rapidly, with heart and kidney disease confining her to a wheel chair.

Jackie Cochran, a friend to four presidents and dozens of European royalty, and one of the most influential and dynamic women of the 20th century died in 1980, at the age of 68. Although the funeral services were private, only fourteen people attended, a poor showing for a person who contributed so much to the advancement of women and aviation.

Helen Richey

Helen Richey was born in McKeesport, Pa. on November 21, 1909. She was a self-proclaimed tomboy who ran away from home at the age of twelve, to join a circus. Her authoritarian father promptly brought her home. Richey preferred male clothing and wore her hair in a boyish bob long before it was stylish. She shunned dolls, preferring to play with mechanical toys. After graduating high school Helen enrolled in Pittsburgh's Carnegie Tech, for a career in education. After a few months she found it dull and dropped out. She stood only 5 feet, 4 inches tall, but those who saw her fly said she was a natural. Ironically, her aviation interest began by accident. Richey lived near an airport and one day she and a friend decided to take a ride to Cleveland just for fun. They flew to Cleveland perched on some mail sacks inside a *Waco* biplane. When Ruth Nichols landed at Cleveland the same day, dressed in her white flying suit, Helen's eyes popped. When she saw the newspapermen, photographers, and autograph seekers gathering around Nichols, Richey suddenly knew what she wanted to do with her life.

Helen Richey, like Earhart, was ahead of her time. She earned her private license in 1930, at the age of 20, and her father, a school superintendent, celebrated with her by giving her an open cockpit biplane. Then she announced the unthinkable: She was going to become a commercial pilot for an airline. She was smart enough to realize that a pilot's job on an airline was in the future. She became instead an aerobatic pilot. That led to a job with Curtiss Wright. By 1932, she had nationwide popularity when she finished third in the Amelia Earhart Trophy Race. In 1933, she was the copilot for Frances Marsalis when they set a 10-day endurance record. Richey herself gained individual prominence in this race. When the hose of the refueling plane ripped the fabric of their plane, Richey climbed out on the fuselage behind the wings and mended the torn fabric. (Fig. 3-4)

In 1934, Richey won the Woman's National Air Meet. In the same race, her friend Frances Marsalis crashed to her death. That left Richey depressed. Richey decided that she needed a more stable flying job and applied to several airlines for a pilot's job. Coincidently, Central Airlines and Pennsylvania Airlines were in stiff competition for the same mail routes, and Helen unwittingly became a pawn. Central Airlines realized there was great publicity value and novelty if they hired a woman as a pilot. Since Richey had already applied to Central for a job, why not hire her? In 1934, Central Airlines broke precedent and hired Richey. Helen Richey became the first woman to pilot a commercial airliner on a regular scheduled route. She made her first flight as a pilot on December 31, 1934, flying a Ford Tri-Motor from Washington to Detroit. The newspapers prematurely hailed the move as breaking new ground for women in aviation and, "the dawn of women coming of age."

Soon Richey became increasingly aware that she was a public relations agent more than a pilot. She found herself on the lecture circuit giving interviews, posing for pictures, handing out autographed souvenir postcards to school children, and not doing much flying. There was also vehement opposition and blatant discrimination from the rest of Central Airlines' pilots. To Helen's disappointment, the male pilots rejected her application for union membership, and because she was a woman, the CAA warned Central Airlines not to let her fly in bad weather. After 10 months she resigned from her job, saying she was not going to be a

Fig. 3-4 Helen Richey

fair-weather pilot. Central Airlines had let her fly only about a dozen round trips. She continued to fly privately and in competition, and later, Amelia Earhart was instrumental in getting her a job in a government sponsored air-marking program.

By 1941, Richey had more than 10,000 hours in her log book, and she became the first woman licensed as an instructor by the newly formed

Civil Aeronautics Authority. In 1942, Richey became the commandant of the American wing of the British Air Transport Auxiliary. She came back to the United States in 1943 and spent 16 months ferrying bombers for the Women's Airforce Service Pilots (WASPs). By the end of the war, Richey held the rank of Major.

After the war, unable to find a flying job, Richey became despondent. With her savings almost gone, she saw only one way out. On January 7, 1947, she made headlines for the last time. Police found her in her bedroom, dead of an apparent suicide. There were too many veteran male pilots and no one wanted a female pilot, - or so she thought.

Ruth Nichols

In the 1930s, a pilot's license was an expensive proposition. With women's annual incomes averaging $800, and a pilot's license costing about $500. It was not surprising that women fliers worked in other fields to earn money to fly.[22] Viola Gentry, who set the first woman's solo endurance record in 1928, had paid for her lessons by working as a cashier in a Brooklyn cafeteria. But, many women in early aviation came from backgrounds of wealth. Ruth Nichols and Laura Ingells, for example, came from fashionable and expensive private schools.

Ruth Rowland Nichols was a socialite and graduate of Wesley University. Her first flight in 1919 was a high school graduation present. By 1923, she soloed in a seaplane and became the first woman licensed in a flying boat. She went on from there to fly every type of aircraft developed. She was rated in the dirigible, glider, autogiro, landplane, seaplane, amphibian, monoplanes, biplanes, triplanes, twin and four engine transports and supersonic jets. Nichols was the first of three women to earn an Air Transport Pilot rating in 1929, and the only woman to hold three different world records simultaneously; women's altitude (28,748 feet), speed (210.5 mph), and non-stop, Oakland to Louisville (19 hrs. 16 min.) between 1931 and 1932.

Nichols attempted to fly the Atlantic, but crashed in Newfoundland, and seriously injured her back. This injury proved yet another challenge for Nichols and she flew for most of her remaining life with a steel back brace. Her injury gave her an idea to organize the Relief Wings, a flying ambulance for mercy missions. This group later became part of the Civil Air Patrol.

Admiral Richard E. Byrd named Ruth Nichols and Amelia Earhart as the two most outstanding women pioneers in aviation. She was also the first woman to become director of the multi-million dollar Fairchild Aviation Corporation. Ruth Nichols died in 1960 at the age of 59.

---

[22] Strother, Lt. Col. Dora Dougherty, USAFR. "The W.A.S.P. Training Program." *American Aviation Historical Society Journal.* Vol. 19 Winter 1974

Laura Ingells

Laura Ingells was one of the most distinguished pilots America ever produced, but she did not start out as a flier. After graduating nursing school at the age of 16, Laura studied music and went on the road with a theatrical group as a Spanish dancer. Ingells went on to study in Paris and Vienna. Laura was an accomplished linguist and one of the bravest of the barnstormers. In 1936, she came in second in the Bendix Air Race, the year Louise Thaden and Blanche Noyes were the winners.

After soloing in 1928 with only 13 hours of instruction, she became hooked on aviation. Two years later she became the first woman to graduate from a government-approved aviation school with her commercial license.

In 1930, residents of Muskogee, Ok., promised her a dollar for every loop over the record setting 344 she had flown in St. Louis. Laura flew 980 consecutive loops and her fans wondered if the only way to bring her down would be to shoot her down. The same year, she also set the unbelievable record of 714 barrel rolls in a single flight, exceeding all male and female records.

In 1934, she was the first woman to circumnavigate South America and the first woman to fly over the treacherous Andes Mountains, and if that wasn't enough, in 1935 Ingells broke the records for non-stop east-west and west-east transcontinental flight.

Laura Ingells was a pacifist, and in 1939, she "bombed" the White House with anti-war pamphlets. She landed in prison and after her release in 1943, dropped out of sight. She folded her wings in 1961.

Forty years later, Louise Thaden described women in aviation in the Thirties. "It was the first time women began to be accepted on their own merits as pilots. It was a time of growth and exploration, when all "firsts" were really firsts. It was a time when camaraderie existed because words were not always necessary between fellow pilots, a time of instant friends and a spirit of cooperation, and a sense of something shared." [23]

[23] Claudia Oaks, *United States Women in Aviation 1930-1939*. Smithsonian Studies in Air & Space #6; p.11

Chapter 4

Women and Warbirds

We realized what a spot we were in. We had to deliver the goods, or else there wouldn't ever be another chance for women pilots in any part of the service.

Cornelia Fort - WASP 1942

By 1940, little had changed for women - economically. About one fourth of American women earned their own money - virtually no change from the days of Harriet Quimby. Full time salaries for clerical workers averaged about $1,200 a year, but women's income had not increased much over the previous decade, averaging only about $850 a year. A pilot's license cost about $750. Most young women were expected to marry and for the most part their husbands did not think women should be pilots or be allowed to earn their own money. Only about 15 percent of married women worked. For a wife take flying lessons in 1940, the husband would have had to relinquish half of his annual salary. Those women who were able to break the financial barriers and formal societal structure never returned to the traditions expected of them. Many of the women were of independent means and they played a crucial role during World War II.

Aviation had not witnessed, nor is it ever likely to do so again, the accomplishments of the Women's Airforce Service Pilots (WASP). There was opposition from many sectors to the concept and to the formation of the first experimental group. There was also opposition to their existence from the day they were formalized until the day they were disbanded. There was still opposition to their recognition twenty-five years after the war was over. Despite seemingly overwhelming odds, they succeeded time after time when others predicted their failure.

Three years before the United States entered the war, Jacqueline Cochran suggested to the Army that American women could serve in non-combat flying roles to free the men for emergency war preparedness duty. Part of Cochran's proposal was to initiate a program for their training prior to the outbreak of hostilities. At the time of her proposal there was no manpower problem, so the Army firmly but politely declined her proposal.

Meanwhile, without any knowledge of Cochran's proposal Mrs. Nancy (Harkness) Love, another prominent aviator made a proposal to use a small number of well qualified women pilots for non combat flying duties. Nancy Love wanted her group of twenty-eight women pilots to ferry aircraft under the auspices of the Air Transport Command (ATC).

Fig. 4-1 Nancy
Love

On September 10, 1942, the Army Air Force announced the formation
of the Woman's Auxiliary Ferry Service (WAFS), under the control of
Colonel William Olds, of the ATC, with Mrs. Love as squadron commander. (Fig. 4-1)

Cochran had wanted a larger group trained and assimilated into the
Army Air Force. A survey showed there were only 50 women pilots with
more than 500 hours, 83 with more than 200 and about 2,000 with under
200 hours. Olds recommended that they use the 50 high-timer pilots to
ferry training aircraft. Cochran, disturbed by this, recruited 22 women
and took off for England in early 1942 to fly with the Air Transport
Auxiliary (ATA) of the British Ferry Command.

When the United States entered the war it was on two fronts: Europe
and Asia. Manpower requirements became increasingly critical but
resistance to women pilots entering the active service remained steadfast. However, by late 1942, the need for a large number of trained pilots
became a profound emergency. The government's first reaction to the
growing shortage was to relax the pre-war pilot candidate standards
requiring single marital status and two years of college. Pilot goals in

early 1941 were 30,000 but raised to 50,000 by December. In January 1942, the number rose to 70,000 and by October, 1942, 102,000.[1] All cadets would be drawn from the monthly quota of men allocated to the Army Air Force.

The supply of aviation cadets could not meet the demand, and in November 1942, Congress lowered the draft age to include 18 year old men. In December, 1942, the President ended all voluntary aviation cadet enlistments, forcing all male applicants for flight training to be in the Army before they could apply to the cadets. Jackie Cochran's proposal three years earlier suddenly became the solution to the problem. In September 1942, Cochran came home from England and Roosevelt appointed her Director of the Woman's Flying Training Detachment (WFTD). Cochran's job was to supervise and coordinate the training of the women pilots for assignment to the ATC. Their job was to ferry the new fighters and bombers to air bases throughout the United States, thus freeing the male pilots for combat. In 1943, they merged with the WAFS to form the Women's Airforce Service Pilots with Cochran as Director, and Love as the Executive Commander. The first women pilots called for this duty were the "high timers" with 500 or more hours. These women earned $250 a month and spent 30-40 days in advanced training for navigation, meteorology, and military law. These women were a unique combination of affluence, determination, and patriotism. Few women in those days had the finances to accumulate 500 hours of flying time.

Later the President and Mrs. Roosevelt invited Cochran to lunch to discuss Britain's war. Her contacts with Lord Beaverbrook and Winston Churchill made her an invaluable asset to the President. Mrs. Roosevelt became interested in Cochran's work in England to recruit women pilots and later opened doors for her in Washington. Mrs. Roosevelt was a strong voice for women in aviation, and had always wanted to learn to fly. The Secret Service, however, had always thwarted her efforts.

Cornelia Fort

Cornelia Fort knew she was going to join before the organization even had a name. Later she recounted her front row seat to the attack on Pearl Harbor. "I was never more certain than on the morning of December 7, 1941. At dawn I drove from Waikiki to the John Rodgers civilian airport, next to Pearl Harbor, where I was a civilian pilot instructor. I began to practice take-offs and landings with my student shortly after six-thirty. Just before coming in for the last landing, I looked around and saw a military plane headed right for us. I jerked the controls away from my student and jammed the throttle wide open to pull above the oncoming plane.

"As it passed underneath, I saw red balls on the wings shining in the sun. We were familiar with the emblem of the Rising Sun on passenger

---

[1] Dora Dougherty, Lt. Col. USAFR. "The W.A.S.P. Training Program." *American Aviation Historical Society Journal.* Vol. 19 Winter 1974 p.298.

ships but not on military planes. I looked toward Pearl Harbor and saw billowing black smoke. I still clung to the hope that it might be a coincidence, or maneuvers.

"I looked up and saw the formations of silver bombers riding in. I saw something fall from one of the planes and go glistening down, until it exploded in the middle of the harbor. I knew the air was not the place for my little airplane. I landed quickly and a few seconds later a shadow passed over me, spattering bullets. We counted anxiously as our little civilian planes came flying home to roost. Two never came back. We found them washed ashore, weeks later."[2]

Fort remained on the island for three months before returning to the United States. None of the pilots wanted to leave, but there was no longer any civilian flying. Each had some individual score to settle for the wanton murder and destruction.

"When I got stateside the only way I could fly was as an instructor in a civilian pilot training program. Then came a telegram from the War Department announcing the organization of the Women's Auxiliary Ferry Service and ordering me to report within 24 hours if interested. I left at once.

"Because there were so many disbelievers about the ability of women pilots, officials wanted the best possible qualifications to go with the first group. We realized what a spot we were in. We had to deliver the goods, or else there would never be another chance for women pilots in any part of the service."

Early in the program, a reporter asked Cornelia Fort why she flew. "None of us can put into words why we fly. It is something different for each of us. I can't say exactly why I fly, but I know why as I've never known anything like it in my life.

"I, for one, am profoundly grateful that my one talent, - flying - happens to be of use to my country when needed. That's all the luck I ever hope to have." Cornelia Fort died in 1942 in a mid-air collision while ferrying a bomber.

The WASP Requirements

As the war expanded there was a need to attract more women to this newly formed branch of the war effort. Women who were high school graduates and U.S. citizens between the ages of 18 and 34 were eligible. The initial 500 hour flight time requirement soon became 200, then 100, and finally dropped to 35, when the universe of women trained as civilian pilots decreased. The women had to pass the same physical and mental tests as the men and they drew $172.50 a month. There was a deduction of $15-$20 for quarters, and meals were extra. Courageous women from all walks of life sought to join this group, a group that isolated them from men, restricted their free time and regimented their lives as strongly as

2
   Helen Collins. *Naval Aviation News*. "From Plane Captains to Pilots." July 1977

any military organization did. They were schooled in math, physics, navigation, weather, Morse code, military courtesy, and flight training 16 hours of every 24 hours. After 24 weeks of this rigorous routine they graduated and their salary increased to $250 a month. The trainees received ill-fitting men's overalls and flight clothing and had to provide their own dress uniforms, khaki slacks, white blouse, and caps. Upon graduation in February 1944, the WASP's Santiago Blue uniform became standard issue. (Fig. 4-2)

Despite the opinions that muscle was needed to fly airplanes and

Fig. 4-2 WASP Cmdr Dottie Young watching a graduating class

women did not have the muscular mass, the WASPs were unremarkable in physical characteristics. They averaged five foot four inches, and weighed 128 pounds. The physical training was the same as that given to the male Army Air Force cadets, except in one area. The women objected to the calisthenics, at least to the 25 push-ups, and refused to do them. They said they weren't in training to become muscular circus women, but pilots. With their objection sustained by the Army the WASPs compromised and did ten.

Fifinella

Just as the men had adopted caricature mascots and good luck charms, the women too had their mascot. Walt Disney was so enthusiastic about the work of the WASPs that he created a new character, a little sprite called "Fifinella," the good sister of the "Gremlins." Fifinellas, unlike their ugly brothers, bring good luck to a pilot and "for a whiff of perfume, a Fifinella will practically do your navigating."[3] (Fig. 4-3)

While a WASP's life was free from people shooting directly at her, it was at best a tough assignment requiring energy, stamina and courage. The working day was almost invariably from dawn to sunset. Delivering an airplane might last from two to ten hours or go on for ten days. Often a pilot ate lunch only when a Red Cross truck drove out to the runway and passed a box lunch to her while the plane was refueled.

The WASPs were at first forbidden from flying in bad weather, at night and stunt flying, although they did practice night landings, and ran the engines at low rpm until they were broken in. Pragmatically, they believed a plane crashed before delivery was one less for combat.

When the pilot delivered a plane, she had to get herself back to her home base. WASPs had top priority on airlines but getting to an airline was sometimes a major problem, since the Army Air Force fields were often in the boondocks. On one odyssey a WASP drove fifty-three miles

Fig. 4-3
Fifinella

3
Betty Peckham. *Women in Aviation.* Thomas Nelson & Sons, Inc. 1945; p.6

in an Army staff car to the nearest bus line. After waiting two hours for the bus she spent another six hours on the bus, sitting on a folding chair in the aisle. After that there was another three hour wait in the bus terminal. Then she spent another hour on a bus to the airport where she waited two hours for an airline. After three more hours on the plane she found herself in Washington's National Airport. Seventeen hours had gone by, and she was not yet home. All that remained was a short train ride to Wilmington, Delaware, and a wait for a staff car from the base to pick her up. Then she usually got about seven hours of sleep before the next flight.[4]

Teresa James had been making floral sprays in 1934, and was afraid of heights. A boyfriend introduced her to flying and a year later when he walked out on her, she didn't shed a tear. She had become focused on her career as a aerobatic pilot and specialist in inverted flying tactics at Buffalo's Aeronautics Institute. James, one of the original ATA ferry pilots, was a major in the WASPs. She had one assignment to travel from New Castle, Delaware, to pick up a P-47 at the Republic factory in Farmingdale, Long Island, and fly it to Republic's modification center in Evansville, Indiana. She didn't pack for an extended trip as she expected to be gone just a day. Four grueling weeks later she landed back at New Castle, having criss-crossed the country delivering planes assigned to her at each base where she landed.

Discrimination

Even with the unyielding support of Jackie Cochran and Nancy Love, the women still faced discrimination and prejudice. The women were routinely grounded during their menstrual cycles. Mrs. Love protested the action and appealed to the ATC headquarters. The air surgeon studied the matter and removed the restriction when he found that menstruation did not interfere with their flying.[5]

Women also suffered at the hands of some of their male instructors. In the Fall of 1943, a WASP crashed in North Carolina. When the official report did not arrive, Cochran flew to the base. After wading through bureaucracy and uncovering some misfiled papers, Cochran discovered a report that cited traces of sugar in what was left of the gas tank. A WASP had died but Cochran went no further. The scandal would have ended her program. In later years Cochran admitted other WASPs had died from sabatoge, and or shabby maintenance.[6]

The early WASP training groups did very well primarily because of their high-time flying experience before the war. Some pilots in later groups were classified unsatisfactory after a series of accidents. The ATC blamed Cochran's training program, claiming it was inadequate, and

---

[4] ibid, p.9

[5] Victor Chun. "The Origin of the W.A.S.P.s" *American Aviation Historical Society Journal.* Vol. 14, #4 Winter 1969 p.260

[6] Deborah Douglas. *United States Women in Aviation. 1940-1990* Washington, D.C. Smithsonian Studies in Air and Space. #7 p.51

Fig. 4-4 WASP Teresa James delivers the 10,000 P-47 Thunderbolt

Fig. 4-5 Ann Carl

Cochran countered with the claim of prejudicial attitudes toward women pilots by some male check pilots. Several newly assigned women pilots received unsatisfactory fitness reports and steps were being taken for their discharge. Cochran requested an investigation. The findings showed that there was discrimination in certain ferry commands against women in transition training to larger planes. The investigators found the attitudes and methods in conducting the flight checks were obstructional and unfair. Check pilots were sometimes resentful of women pilots and the program, and wanted its elimination. Reprimands went to the commanding officers of certain ferry training groups for this treatment, and the WASPs in question were retrained and advanced to the extent of their ability.

Tough jobs

One of the dicer jobs for the WASPs was target towing. It was a tough and dangerous job most male military pilots detested and avoided. The women eagerly volunteered for this assignment, moving quickly into the target-towing business with a limited amount of training and no experience. The women proved not only capable of the nerve-racking job but Army officials discovered the women were better adapted at it than male pilots returning from combat.

The job literally involved being shot at. A WASP would fly at 10,000 feet towing a brightly colored sleeve shaped muslin target on a 2,500 foot tether. Student anti-aircraft gunners would then fire live 90 mm shells at the target. Gunners would sometimes take too much lead on a target and the WASPs would land their planes and discover the rear fuselage and tail section peppered with holes. One WASP had a bullet lodge in her engine, and as she was coming in for a landing, the engine quit, resulting in a fatal crash. Miraculously, no other injuries to the pilots occurred during those exercises, nor were any planes lost, in spite of many close calls.

Besides ferry training, the WASPs participated in four other programs. The "C-60" was a glider-tow training course which yielded disappointing results. The women lacked the physical strength for this fatiguing type of flying. They did, however, perform tasks that added immeasurable contributions to the war effort. Under Top-Secret conditions a group of women were trained to fly radio-controlled target planes. During that training, one WASP sat in a tiny PQ-8 aircraft controlled by a second WASP using radio controls in a "mother ship." The "captive" WASP rode helplessly as her tiny plane zoomed and dived and maneuvered wildly through the skies. In an emergency the WASP in the PQ-8 could override the controls and take control of the aircraft. Other WASPs flew dangerous low-level missions, laying down smoke screens in mock chemical warfare. Some flew fighter aircraft as targets for other male pilots in mock dogfights. The male pilots shot film instead of real bullets.

Transition training to B-26 bombers was more successful, 57 women reported for training, and 29 graduated. The instrument flying program was highly successful; 234 of 246 women graduated the course.

Fig. 4-6 Elizabeth Gardner in the cockpit of a B-26 bomber

Fig. 4-7 WASP V.M. Saunders in front of her B-26 bomber

The WASPs proved step by step that they could fly anything the Army Air Force had, and do it well. Like the men, these women learned cross country navigational skills, practiced flying solely by instrument reference and criss-crossed the nation making deliveries of every war plane in the United States inventory. By December, 1944 the date when all WASPs left the service, the WASPs had been flying every P-47 "Thunderbolt" rolling off factory lines. (Fig. 4-4).

The WASP's successes in non-ferry assignments led the Army Air Force to take a broader view of their capabilities. Consequently, the women received a variety of assignments. They first became instructors but there was resentment toward them from the male instructors and even from some students. They were then given assignments from engineering test pilots in day and night missions to training radar and searchlight trackers. In all these assignments they conducted themselves professionally and did well. The WASPs flew cargo, Top-Secret weapons and personnel, and had the dangerous job of testing the new planes to be certain they were safe for use by instructors and students. They were also some of the first pilots to fly the experimental jets. In October, 1943, Ann Baumgartner Carl(Fig. 4-5) flew the YP-59A and became the first woman to fly a jet. Whatever their assignment, the WASPs performed with a willingness and expertise that earned them acceptance and praise. As in the previous decade, the women were also used to demonstrate airplanes with fearsome killer reputations, like the B-26 Marauder, often called the "Widow Maker." If a woman could fly the plane, it must be safe, was the subtle implication. (Fig. 4-6, 4-7)

Make Us Real or Make Us Gone

By 1944, combat losses were far below predictions and large numbers of Army Air Force pilots were rotating home to take over stateside duties. Throughout the war several efforts were made to assimilate the WASPs into the military establishment. The last bill introduced in September, 1943, and revised in February, 1944, looked promising. Some previous efforts to assimilate the WASPs had been adamantly opposed by Cochran. Congress was willing to merge them into the Woman's Army Corps (WACs) but Cochran said bluntly, "Over my dead body."[7] She would have no part of an organization that had no knowledge of flying.

She stood her ground that the WASPs should be commissioned directly into the Army Air Force. The Army Air Force launched their own secret investigation and recommended maintaining the status quo. In their opinion the civilian status of the women offered greater flexibility. General Arnold wanted to keep the WASPs in service, but all previous efforts had failed in the Senate.

Strong lobbying by civilian pilots under Army contract had prevented assimilation of the WASPs into the Army Air Force. The civilian pilots believed they would lose ground in seniority and jobs if the women

[7] Jacqueline Cochran, Maryann Bucknum Brinley: *Jackie Cochran The Autobiography of the Greatest Woman Pilot in Aviation History.* New York Bantam Books 1987;

Fig. 4-8 WASP June Elington in an A-24 fighter

Fig. 4-9 WASP Mary Ball on the way to the flightline

became part of the Army. The Army was closing down the pilot training program and the 10,000 highly paid civilian instructors would be out of jobs and eligible for the draft or unemployment lines. A bitter battle followed. In August, 1944, Cochran made another appeal. She said, in effect, militarize or disband the WASPs. The queen WASP had not lost her sting. The newspapers splashed Cochran's statement across the front pages and many called it an ultimatum. Her poor timing and quick temper had gotten the best of her. Victory in Europe was imminent and the need for combat pilots had dropped drastically. When an 11th hour effort in the Senate failed, General Arnold, who had gone on record saying, "Women can fly as well as men" had to announce the WASP program would end on December 20, 1944.

The WASP Record

In April 1944, when passage of a Bill in Congress (HR 4219) to militarize the women into the Army Air Force seemed imminent, 460 women enrolled in a three week Basic Military Training Course for Officers, at the School of Applied Tactics, in Orlando, Florida. All the women enrolled passed the course, but the school closed on September 29, 1944, when it became obvious that the deactivation of the WASPs was imminent.

The WASPs were at their peak efficiency in December, 1944, and had compiled an impressive record. They delivered 12,650 planes of 77 different types. Fifty percent of all the high speed pursuit planes ferried in the United States were flown by the WASPs. More than 25,000 women applied to the WASPs and of the 1,830 women admitted, 1,074 graduated, and they flew more than 60 million miles. Thirty eight lost their lives in accidents, eleven in training and 27 in operations.[8] They did not have military insurance since they were not part of the military and no civilian company would insure them so the women were buried without military honors.

Jacqueline Cochran, with some bittersweet emotions, addressed the last WASP graduating class on December 7, 1944, and summed up the accomplishments of the program. "Happiness also swells within me from the knowledge that the WASPs have successfully completed their twofold mission. We have flown scores of millions of miles in relieving the pilot shortage and we proved that women can be trained as pilots easily and used in many ways in the air effectively. What the WASPs have done is without precedent in the history of the world.[9]

[8]   Valerie Moolman. *Women Aloft*. Time Life Books, VA. 1981, p.153 and 30th Memorial WASP reunion.

[9]   Dora Dougherty, Lt. Col. USAFR. "The W.A.S.P. Training Program." *American Aviation Historical Society Journal* Vol. 19, Winter 1974; p.206

"You have freed male pilots for other work, but now the war situation has come when your volunteered services are no longer needed. The situation is that, if you continue in service, you will be replacing instead of releasing our young men. I know that the WASPs wouldn't want that. So, I have directed that the WASP program be deactivated.[10]

In closing, Cochran said, "After the war is over, there will be a tremendous growth in our civilian airline network. All this increase in air travel will require the building of many more airports. At these airports will be girls employed in radio and control tower work, and girls in machine shops learning aircraft and engine maintenance."[11] There was no mention, however, of them working as airline pilots. It was unthinkable to consider them to fly as pilots on scheduled airlines. (Note: Since they did not have veteran's status, the airlines literally ignored their presence and pilot skills.) WASP Beatrice Hayden (Bee) said, "When I got notice that we were being disbanded, I sent letters to aircraft manufacturers, airports, airlines; anyone I could think of to try and get a job in aviation. I never saw 'no' written in so many different ways."[12]

The WASP program of World War II showed the psychological response of women who participated in high-performance flight activities, including combat missions, to be positive. In fact, morale was so high that it was necessary to discourage the women from "stunt" flying. The women were not overly cautious and, on the contrary, were always enthusiastic inspite of the many "man-made" difficulties they faced.

The WASP record was outstanding. Their safety record was better than male pilots flying similar missions. The WASP accident rate was .001% while their male counterparts in the United States had a .007% accident rate.[13] The WASP fatality rate was also lower with .060% per 1,000 miles compared to .062% for the Men.[14] They lost less time for reasons of physical disability than did their male colleagues. Sources suggest that their lower lost time is the result of less drinking and more dedication, juxtaposed by the propensity of male pilots to drink to excess and travel with "a little black book."

The WASPs were assigned to 134 bases throughout the United States, and served in Second and Third Air Forces, Material Command, Weather Wing, and Flying Training Commands.

10   Jacqueline Cochran, "*Final Report of the Women Air Force Service Pilot Program*," manuscript National Archives: Washington, D.C., 1945.

11   ibid p.154

12   Stuart Leuthner; Oliver Jensen. *High Honor*. Washington, D.C. Smithsonian Press 1989, p.296.

13   *Flying*. December 1944.

14   *Flying*. January 1944.

Fig. 4-10 The WASP's B-29 bomber "Ladybird." Col. Paul
Tibbets, Jr., (l) WASPs Dorothea Johnson & Dora Dougherty
Strother, with the crew of "Ladybird."

Fig. 4-11 (T-B), WASPs Dora Dougherty Strother
Joyce Sherwood, and Florence Knight on a Curtiss "Helldiver"

Fig. 4-12 Nancy Love (l) Cornelia Fort, Mary Clark, Teresa
James, and Betty Gilles, four of the original 27 WAFS

It is ironic to note that after racking up an enviable record, one writer
said, "Most of the members were immediately offered excellent positions
in civil aviation. Some served as instructors, but none flew for the leading
airlines."[15] That would take three more decades.

Finally - Recognition
In the mid 1970s, newspapers announced that the U.S. Air Force
planned to train its "first women military pilots." To the WASPs the news
was an outrage and an insult. They began a campaign to gain the recognition long overdue.
In 1977, Senator Barry Goldwater introduced Senate Bill S.247, He
called it the "Year of the WASP," and said the bill was, "To provide
recognition to the Women's Airforce Service Pilots for service to their
country during World War II..." Barry Goldwater was an ideal champion
of their cause. He was a Major General in the Air Force Reserve, and
during World War II had flown wingtip-to-wingtip with the WASPs as a
pilot with the Air Transport Command. At the time, he had 12,000 hours

---

[15] Betty Peckham. *Women in Aviation.* Thomas Nelson & Sons, Inc. N.Y. 1945;
p.10

in the air in over 200 different aircraft. He was Congress' most experienced pilot and his voice would silence any opposition from the Senate. For the House battle he enlisted the aid of Colonel W. Bruce Arnold, the son of the late General Hap Arnold. Bruce Arnold was also a strong supporter of Cochran and her WASPs. Arnold in turn enlisted the aid of Congresswoman Lindy Boggs, and Antonia Clayes, assistant secretary to the Air Force. The battle lines were soon drawn, the old guard did not want to give up their age-old philosophy. Arnold and Boggs had one ray of hope in the House Veterans Affairs Committee to get their bill passed. The second ranking minority member was the committee's only woman, Margaret Heckler, of Massachusetts. A Republican, Heckler was co-chair with Democrat, Elizabeth Holtzman, of New York, of the newly formed Congressional Women's Caucus. In March, through Heckler's efforts, the WASP bill became the only piece of legislation to be co-sponsored by every woman member of Congress.

By May, the battle had just been joined, however. The Senate hearings took the full brunt of the "old guard." The American Legion testified in the Senate, "In the history of our nation, the veteran has, from the time of the Revolution occupied a special place. It is highly prized and valuable, and is to be shared only by those who have earned it. To legislate such a grant of benefits would denigrate the term "veteran" so that it would never again have the value that it presently attaches to it."[16] The Legionnaires even threatened to remain forever absent from discussions that advocated veteran's causes should the WASP Bill pass.

The old animosities toward female pilots, and some say Jacqueline Cochran's ability to make long lasting enemies, died slowly. The "old guard," however, had lost its punch. On November 23, 1977, the eve of Thanksgiving, President Jimmy Carter quietly signed the WASP Bill and it became law of the land - more than 30 years after the WASPs had been disbanded. Official acceptance into the Air Force did not come until 1979. That year, the Air Force accepted them as part of itself. Cochran had lived to see justice. The WASPs received the World War II Victory Medal and American Theater Medal in 1984.

The media reporting of the WASP Bill in-fighting had a positive effect on women in aviation. It prompted young women to investigate general aviation as a career alternative. In 1970 there were 29,832 women licensed as pilots. Nine years later there were 52,392 women licensed, or an 80% increase. In the same time period, the men only saw an 11% increase.[17]

16   Congressional Record

17   Deborah Douglas. *United States Women in Aviation. 1940-1990.* Washington, D.C.
      Smithsonian Studies in Air and Space. #7, p.103

Fig. 4-13 WAFS pilot Barbara Erickson (l) & WASP Betty Tackaberry marching with color guard.

Fig. 4-14 WASP Holly Hollinger in cockpit of AT-6

Chapter 5

The Whirly-Girls

We fly the Chinooks with supplies to the troops. As aviators and soldiers. This is the moment we have trained for. There is a man flying in front of me and one behind. We women are professional - we're soldiers - not men and women.

Major Marie Rossi

The autogiro, something of a cross between a conventional fixed wing plane and a helicopter, was first developed in the 1920s. It was limited, however, in that it could not take off straight up from the ground but needed a short roll. It could not fly backwards or sideways either. It was, however, a precursor of a new form of flight, and women were involved with this new vehicle too. Amelia Earhart, one of the first women to fly an autogiro in 1931, flew it to an unofficial altitude record of 18,451 feet. She then flew it across the country making her the first woman to accomplish a transcontinental flight in the autogiro.

The first true vertical flight machine, a Focke-Achgelis helicopter, was test flown in 1938, by Germany's foremost pilot, Hanna Reitsch (Fig. 5-1), but World War II halted its development as a commercial vehicle. During the war, however, research into this new form of flight continued and the Army used some Sikorsky YR-4 helicopters in Burma and England. When the war ended, the helicopter was ready for commercial application, and women were ready to make use of the new air vehicle.

The first woman in the United States to earn a helicopter rating was former WASP, Ann Shaw Carter of Fairfield, Conn. (Fig. 5-2) She is Whirly-Girl # 2, (Hanna Reitsch was #1). In June 1947, Carter soloed a Bell Model-47 helicopter at New York's Westchester Airport. She was hired soon after as a pilot for the Metropolitan Aviation Corp., one of the first passenger helicopter services in the United States. Carter also became the world's first female commercial helicopter pilot. Close behind her were, Nancy Miller Livingston, Priscella Handy Swenson, and Marilyn Grover Heard, all of whom earned their helicopter ratings before the end of 1947. Livingston (Fig. 5-3) also flew for three years in WW II, with the British Ferry Service. She later became the first female crop dusting pilot in America. (Californian Elynor (Rudnick) Falk, introduced the crop dusting helicopter to New Zealand. She was also on the team of helicopter pilots that helped map Alaska.)

Fig. 5-1 Hanna
Reitsch

Fig. 5-2 Ann Shaw Carter

Jean Ross Howard

The Whirly-Girls is an international organization of women helicopter pilots that provides a support network for women who fly helicopters. It promotes the exchange of helicopter information and advances the acceptance of rotary wing aircraft. Washington, D.C. native, Jean Ross Howard, came up with the idea. Howard learned to fly at George Washington University, in a government sponsored Civilian Pilot Training Program. "It was the best college course I ever attended," she'll tell you smiling broadly. (Fig. 5-4) Jean is a former Civil Air Patrol pilot and World War II WASP trainee. Today she serves as the Whirly-Girls' Chairman of the Board.

It began back in 1953 at the International Air Pioneers Dinner in Washington, D.C. The biggest names in Aviation were there including the late Lawrence Bell, President of Bell Aircraft Corporation, producer of the first commercial helicopter. "In talking with him, I once again dropped a "hint" that I'd like to learn to fly a helicopter." Bell had taught French test pilot Jacqueline Auroil to fly his machine, and Howard felt she'd be better qualified in her job with the Helicopter Council of the Aircraft Industries Association. "Then too," she said, "it was love at first flight when I'd been a helicopter passenger a few months earlier.

"Mr. Bell must have heard my "hint" before. He said, "All right, Jean, get yourself down to the Bell plant in Fort Worth, and we'll give you the same course we gave Madame Auriol." Eighteen days, later Jean Howard was ready for her flight test. She passed the test, and when she landed her instructor said, "Well Jean, you're a real whirly girl now." The name stuck.

"I began to wonder how many other American women had helicopter ratings," said Howard, "and when I found out I was the eighth woman accredited to fly helicopters, I modestly proclaimed myself to be the eighth wonder of the world." Then Howard decided to verify how many women in the world flew helicopters. It turned out twelve other women had been flying helicopters for years. "I was 'lucky 13' and after I found out there were thirteen of us I thought we should get together."

The Club

The Whirly-Girls started in 1955 as an unofficial group with no officers, no dues, and no records of meetings. Soon it became clear that if they were to stay together and in touch, they needed a recording secretary. The choice was obvious, Jean Howard. They even adopted a logo, a little caricature of a helicopter with long eye lashes and a Betty Boop expression. "From the beginning our instant rapport was amazing, and it still continues today," said Howard,

By 1962, the organization had grown to 40 women world-wide and nine had qualified as helicopter flight instructors. Twenty years later there were 87. In August, 1991, there were 781 Whirly-Girls from 26 countries. The membership requirement is a certified helicopter rating from the U.S. Federal Aviation Administration or its foreign equivalent. The Whirly-Girls even have a language all their own, shared exclusively in the helicopter community: cyclic control, collective pitch, tail rotors, torque forces, and autorotation. They don't hold meetings, they have "hoverings."

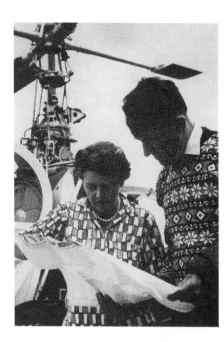

fIG. 5-3 Nancy Miller Livingston with Sir Edmund Hillary

The Record

The early helicopters were as difficult to fly as the early fixed wing aircraft. The machines were, however, technically complicated and difficult to handle. Former WASP DorothyYoung was the first woman in the free world to hold an airline transport rating for helicopters and she had a major disadvantage, weighing only 97 pounds. Helicopters, like the early monoplanes, also have weight-load ratios to maintain. When she didn't have a passenger aboard, Young carried bowling balls aloft as cockpit ballast to distribute the weight properly. Other Whirly-Girls carry sandbags, or a stuffed toy filled with lead.

Until 1961, all helicopter records held by women were earned by Russian women. Dr. Dora (Dougherty) Strother, a past president of the Whirly-Girls and a Ph.D. in human factors with the Bell Helicopter Company, changed that by bringing the records home to America. With only 34 hours flight time in a helicopter Dr. Strother went straight up to 19,406 feet to break the Russian's record. She promptly set another record by flying in a straight line for 404.36 miles. Dr. Strother was no stranger to aviation and like most helicopter pilots could fly many different types of aircraft. Her distinction was that she was a former WASP and was one of two women qualified to fly the B-29 bomber.

In 1962, Gloria Jean Miller, daughter of Whirly-Girl Katherine and Russ Miller, made aviation history. On her 16th birthday she soloed in both an airplane and helicopter. A year later, when she became of age - aviation age - she received her private pilot and helicopter ratings. Then it was back to the classroom, but this time, for her driver's license.

In 1965, Gay Maher, a helicopter flight instructor was the first (man or woman) to solo a helicopter coast-to-coast. This mother of three was in the air 35 hours during the 3,000-mile, 10-day, 80-mile-per-hour flight. She made 33 fuel stops in her Hughes Model 300.

Whirly-Girl Jerrie Cobb soloed after 83 minutes of dual instruction, something that usually takes eight hours. Jerrie got her Private Pilot rating at 16, commercial rating at 18 and instrument rating at 21. In 1959 she was named Pilot of the Year, by the National Pilot Association. She also earned the Gold Wings of the Federation Aeronautique, and the Amelia Earhart Award.

Another Whirly-Girl, a retired school teacher, Alice Weisendanger, decided at the age of 64, that a helicopter would be a great way to get back and forth to her newly acquired semi-precious gem mine. She went out, bought a Hughes 269-A helicopter, and then earned her helicopter rating.

One Whirly-Girl, K.C. Nichols, made it possible for women to fly stunt doubles in parts that had woman pilot characters. Her lobbying efforts with the Screen Actors Guild helped eliminate the "man pilot in the wig" role.

In 1987, seventeen year old high school senior, Kim Darst, of Blairstown, New Jersey set a record also. She became the first person to fly into her high school graduation ceremonies in a helicopter. She landed in a soccer field, changed her shoes, put on her cap and gown and received her diploma. (Fig. 5-5)

Former flight attendant Marlene Morris and her husband have their own "His" and "Her" helicopters. Marlene says she never gets into anything that moves without something to eat, or drink. These unique flying experiences have resulted in her writing, "The Flying Gourmet,"

Fig. 5-4 Jean Ross Howard

Fig. 5-5 Kim Darst at her high school graduation

Fig. 5-6 Jean
Tinsley

just *plane* good recipes for "Wind Sock Salad", "Chopper Slaw" and "Scared Stiff Merinque." This culinary artist says her experiences are limited to, "thirty years of taste and error."

In 1988, Whirly-Girl, Enid C. Kasper, did a scientific survey of women helicopter pilots. The survey was unique in that until then there was no data from any source including the FAA, and Ninety-Nines, on the psychological and physiological profile of a woman helicopter pilot. Kasper's results showed that; the average age was 42, and 42% were married. Ninety-two percent were Anglo, with 41% college graduates and 20% of them had a Masters Degree.

Forty-seven percent were first born children, and most had received their first pilot's license between 17 and 27, although 7% had gotten their helicopter rating after the age of 40.

The discipline of flying had helped 53% of them "absolutely" in other areas of their lives and 47% wanted to fly because of a "profound interest." Another 21% found enjoyment in the "continually challenging environment."

Not surprising 46% of the Whirly-Girls said they encountered sexism "on occasion" while almost 20% said "often." A few even said they "could write a book about it." Only 17% of the women's non-flying male friends found their work "intimidating" and 78% of their male friends were "supportive" and "fascinated."

The survey found some common characteristics among the women. They are non-conformists and free thinkers, and about 50% of them "almost always or frequently" have some fear of heights other than flying. Forty-nine percent have a "positive self image" and feel "in control." Many even describe flying a helicopter as, "a view above life - a separation from all that is trivial - an art - a physical expression unique to a woman's world."

When asked for some common traits she saw among the Whirly-Girls Jean Howard said, "They are doers! They are adventuresome and thoroughly professional."

Howard was once asked why so many woman are attracted to helicopters. She answered this way, "It may be something in our nature. A helicopter stops and then lands. An airplane lands and then stops. In a helicopter, unlike an airplane, you can start to land or take off and then stop; you can back up, and look again or just park (hover) in the air. It has been said we sometimes change our minds. With a helicopter we can, and safely."

Many Whirly-Girls talk about the helicopter experience as "love at first flight," and perhaps Hanna Reitsch best described the feeling. "Potent, yet gentle, like some seductive wine, the fever of flying descended on me coursing through me to my very fingertips."

Jean Tinsley

Jean Tinsley used to think she was born twenty years too soon. One of her childhood dreams was to be an airline pilot. In 1951, she applied to Pan American Airways for a pilot's job. At the time, she was fluent in Spanish, and had an impressive log book of multi-engine time. The Pan American Airlines interviewer didn't get past her multi-lingual skills.

Without hesitation he offered her a job as a stewardess on one of their South American routes.Her reply was thank you, but no thank you.

Although there were no pilots in her family, Jean's first recollection of an interest in flying goes back to taking her five-cents-a-week allowance and going to the store to buy balsa wood airplane models, the kind you build yourself. She also remembers not being interested in dolls and eschewing the other traditional toys. Instead, she preferred to play with put-together and mechanical toys.

At the end of World War II, her father had some left-over gas rationing coupons. He converted them into an airplane ride for himself and Jean. After the ride Jean knew what she wanted to do - fly.

She proceeded to get her private and commercial ratings and in 1961, earned her balloon rating. She has successfully combined a career in aviation that spans more than 40 years, 25 of those years flying helicopters, with raising six children. But it all really came together for Jean in 1965. That was the year she met Jean Ross Howard. Jean invited Jean to a "hovering." At the get-together Jean Tinsley met such legends as Edna Gardner Whyte, Whirly-Girl #10, and Blanche Noyes. When Jean Tinsley left the "hovering," her friend, Whirly-Girl Jackie Waide said Jean no longer had eyeballs. They were rotors. It wasn't long after that Jean Tinsley added her name to the Whirly-Girl roster becoming #118.

When air forces from all over the world met for the second World Helicopter Championship (WHC) competition in 1973, the United States fielded the first, and only all-woman team to that meet. Tinsley was one of the six pilots on the team. Shortly after that, she became the first woman in the world to earn a rating in the constant speed giroplane. Tinsley went on over the years to work through the various levels of judging until 1989, when she became the WHC's chief judge. (Fig. 5-6)

Jean Tinsley recently joined an elite group, and marked another first for women in aviation, but nothing new for Tinsley, she's been the first in many aviation milestones. Her most recent coup was to become the first woman in the world to fly the experimental XV-15 tilt-rotor aircraft. Only two are in existence and they have only been flown by Bell Helicopter test pilots and invited guest pilots like senator and astronaut John Glenn. It is an exclusive group and Tinsley wanted to join. Tilt-rotor pilots have set six world records, and Jean Tinsley decided to set one more record. With a combination of reputation, pilot skills an impressive log book and a winning personality, Tinsley convinced officials at Bell to invite her to fly the new machine. She will tell you that it wasn't quite that easy. When she approached Jack Horner, president of Bell, in 1990, and mentioned that she would like to fly the tilt-rotor, Horner laughed and said that no woman had ever tried it before. Her answer to him was an icy glare. That look broke the ice. On April 23, 1990, she became the first woman to fly this combination helicopter-turbo-prop airplane. Only one thing about the flight concerned Tinsley. This self-described "four cushion pilot" (she's 5'2" and weighs 105 pounds) was anxious about fitting into the cockpit. Happily she discovered she didn't need a cushion, but she still needed a teddy bear stuffed with lead shot as ballast to bring the cockpit minimum weight up to 150 pounds.

After the forty minute flight she remarked, I was flying the aircraft of the future, one my grand children will be flying." Jean, who also has eight grandchildren is now looking forward to flying the newly developed, and larger V-22. When asked if she intends to try it too, she replied, "I'm not going to try it, I'm going to fly it."

Tinsley is co-founder of the Helicopter Club of America and currently its president, and when she found out the sub-title of this book she laughed and said, "I've been living the story for almost 40 years." When asked to elaborate she replied, "Every time I tell someone who is not in the aviation community I fly helicopters they say, 'You fly WHAT?' And when I tell them I'm the first female president of the Helicopter club of America they say, 'You're the president of WHAT?' It really takes people by surprise."

She'd like to see more women in helicopter aviation and advises young women looking into careers in aviation, "Go for it! All the doors have been opened, and you get the same pay as the men for the same work."

Teresa McIntosh

Terri McIntosh recently became a Whirly-Girl. She kept focused on this goal for over seven years but finances limited her aviation education to fixed wing aircraft. "It started out as a long-term goal which seemed like an impossible dream."

"My dream was to become a helicopter pilot and fly for the Los Angeles Police Department (LAPD) Air Support Unit. I discovered that the minimum requirements to apply for that position were five years on the job with the LAPD and a commercial rating in aviation. To someone with my background (no prior knowledge of either subject, and no friends or family even remotely involved with aviation or policework), those requirements could have seemed beyond my drive, interest or educational levels. I could have been discouraged but I said, 'I could do that and I would truly enjoy doing that.' I know I had a tremendous uphill battle of work ahead of me but the end result would make it all worth the effort.' (Fig. 5-7)

"I immediately began testing for a patrol position with LAPD and continued investigating the air support unit. LAPD hired me and I completed over five years in patrol divisions. I then discovered air support had an opening for an observer position, so I interviewed and they accepted me. There are many aspects to the observer position and for that reason it takes a year or so to be well-versed. I trained by day to become a good observer, and my time off I spent training for my fixed wing commercial rating. After I earned my commercial rating, air support began interviewing for the command pilot position.

"LAPD Air Support has an in house training program that is full time and lasts six months and the interview board makes their pilot selection decisions based on many factors and qualifications of the applicants.

"The LA area is comprised of complicated airspace, congestion, low visibility and some of the fastest talking ATCs in the world. LAPD pilots must deal with a high level of stress and must maintain exemplary behavior and an attitude of safety at all times. Noise is a serious issue and

Fig. 5-7 Terri McIntosh

Fig. 5-8 Charmienne Hughes doing a pre-flight check

for these and many other reasons the LAPD command pilot position has very special boots to fill.

Teresa was selected for the position, and through the LAPD training and FAA testing she received her private and commercial helicopter ratings. On January 25, 1991, the captain of the division pinned command pilot wings on Teresa McIntosh. She made her dream come true.

Charmienne Hughes

Three years after she had her first flight, Charmienne Hughes is now president and chief flight instructor of Triad Aviation in Westminster, Maryland. Charmienne first discovered the world of helicopters on a heli-hiking trip with her mother in the Canadian Rockies. Heli-hiking is where a helicopter drops you on the side of a mountain and you hike the mountain. Later the helicopter returns to fly you to another mountain. She'll tell you that it was "love at first flight."

Charmienne decided to get her fixed wing rating first while working a full time job. "I met Jean Howard at Phil and Lee (#426) Hixon's house shortly before my private check ride. Jean has been an absolutely wonderful friend and supporter making me feel welcome even before I became a Whirly-Girl. At the hovering she took the time to reassure me about the check ride which I was convinced I'd never pass. Her enthusiasm is contagious."

Charmienne soon realized that if she were going to be serious about helicopters she would have to devote full time to training. So she did what anyone else with a vision might do. She quit her job and in March 1988, she took her first helicopter lesson. A year later she had her Certified Flight Instructor's rating, and a job. The following year she pulled up stakes, and relocated so she could take a job as the chief flight instructor at a school in Maryland. Unfortunately, the school closed because of finances just four months later. Now Charmienne had to do some soul-searching, and career planning. Since she wanted to continue instructing, she decided she could do that and control her financial future herself if she opened her own school. Today, ten months later, Triad Aviation is a strong and productive company. She has had to hire two instructors to help her with the growing company. (Fig. 5-8)

Leigh Herrmann

Leigh Herrmann has been flying since March, 1989, but aviation became part of Leigh's life through an osmosis-like effect. While she had an uncle who flew in World War II, some of her friends and people she worked with flew so aviation seeped slowly into her being. These days Leigh is working at radio station KTAR, in Arizona, and is half way to her commercial helicopter and CFI ratings. She credits two Whirly-Girls, Diana Stuart, and Roseann Ballard with providing inspiration and being true mentors. (Fig. 5-9)

Leigh would like to see more women earning their living in helicopter aviation, and urges women who are thinking about it, "Not to lay the idea down. It is important for us to succeed in this field. We have an image-shaping role we can play." Leigh feels that, "If you have the strength to start in this field, you must find the strength to hold on to your goal."

Fig. 5-9 Leigh Herrmann

Leigh also finds helicoptering an almost spiritual experience. "Some of my first solo flights were very awakening for me. When you fly out here, it's sometimes 120 degrees and very windy. You get to know yourself very well, up there with no one but yourself to rely on."

Variety

The Whirly-Girls have included (and still do include) doctors, engineers, airline pilots (there are more than 20), the wife of a senator, test pilots, a concert pianist, grandmothers, traffic reporters and homemakers, a true representation across the spectrum of American women, and yet it is a unique club. Today's women can be found employed in a variety of fields that use the helicopter for their livelihood. Perhaps no other field of aviation offers such diversity. Women fly search and rescue missions, air taxi, med-evacs and resupply offshore oil platforms. Others do sales, tours, and perform wildlife control including such diverse activities as elk herding and otter transplants. They fly civilian law enforcement jobs with county sheriffs, and other activities like fire suppression, agricultural, and photographic services are also common.

Pat and Dick Jenkins live on the family's 100,000 acre ranch. Each morning Pat and her bright yellow Hughes 300 helicopter decorated like Woodstock take off to herd 2,300 head of cattle 65 miles out on the ranch. She flies fence patrol, hauls tons of salt-lick blocks, and sometimes acts as a flying chuck wagon for the range riders. Sometimes she ferries them to distant parts of the ranch, saving hours of dusty trail riding. Helicopter range riding reduces the Jenkins' management costs about 40 percent.

Barbara Collins has been flying a police helicopter on patrol in metropolitan Washington, DC for over six years. She's been on the force for 18 years but all her flight training was at her expense. She had to spend five years as an aerial observer, before she could begin the one-year qualification to become a pilot. She covers robberies, drug busts, conducts search and rescue missions and is currently training new police pilots in the department's program she was instrumental in creating.

Bonnie Wilkens has a masters degree in Bio-Aeronautics, the study of the use of aircraft in agricultural work. To complete her Masters Degree, she needed to have six weeks of practical application, so it was off to AgRotors where she tried the helicopter. As the Whirly-Girls say, it was "love at first flight." After completing her Masters program, she returned to AgRotors for her helicopter and flight instructor's ratings. After 1,000 flight hours she began doing agricultural spraying and fire suppression work - Heli-tack missions. As the only civilian woman Heli-tack pilot she operates a Bambi Bucket with 100 gallons of water. From a hover, over a fire, she can dump the water directly on the fire, then refill the bucket by flying over a lake or swimming pool. (Fig. 5-11)

Fig. 5-10 Executive Director Colleen Neivus

Kay Bowman wanted to learn to fly helicopters but needed a job to pay for her lessons. When Kay heard that Jane Fonda needed a cook for her health Spa, Kay applied. Instead of sending a resume, she prepared a whole meal. Her inventiveness landed her the job, and Kay began saving her money for helicopter lessons. Kay went on to win a Whirly-Girl scholarships, and now has her flight instructor's rating. Kay has passed on her skill, by teaching several new Whirly-Girls. Kay is now on her way to opening her own flight school, in Santa Barbara.

In New York City, a woman's voice used to be rare on the radio frequency requesting clearance to the 60th Street heliport. Today it is routine to hear General Electric's Captain Diane Dowd, AT&T's Kathleen Mucha, Captain Cindy Wilson, of Phillip Morris, or Island Helicopter's Caroline Lachmann, on the airways.

The first scheduled helicopter airline pilot was Nadine Fetsko, who flew for Pan AM between the New York area airports and the 60th Street heliport. Her copilot was Whirly-Girl Josie Edmondson.

Some Whirly-Girls have formed their own companies. Barbara Ohliger, of Newark, Delaware is president and chief flight instructor of Sundance Helicopters. Barbara already had a RN degree and two Masters degrees and was the mother of seven children when she turned to helicopters.

Each year the Whirly-Girls, assisted by their Men's Auxiliary, award two scholarships amounting to $10,000 to deserving young women to obtain their initial or advanced helicopter ratings. In June 1991, the Whirly-Girls announced a new $5,000 Major Marie Rossi Memorial Scholarship Fund. Major Rossi was the first woman pilot killed in Desert Storm.

The 1990s will see continued growth of the aviation sciences, and since many Whirly-Girls are employed in the industry you can count on the Whirly-Girls to be on the leading edge of growth in the helicopter industry. That growth will be accelerated by their own rapidly growing membership. In the first six months of 1991, the Whirly-Girls welcomed 25 new members.

Fig. 5-11 Bonnie Wilkens

Chapter 6

Military Aviation

Federal law prohibits women... from directly engaging in aerial combat,... delivery of munitions or other destructive material against an enemy and duties where enemy fire is expected and where risk of capture is substantial.
Title 10 USC 8549

On May 7, 1429, 4,000 French soldiers stormed a fortified British position that was the key to Britain's siege of Orleans. The commander of the assault was seriously wounded by an arrow but continued to lead one assault after another. By nightfall, the French had overrun the fortification, and the British blockade was broken. The leader of this valiant army was a seventeen year old woman named Joan of Arc. Joan perhaps played a more active combat role than many of her sisters.

For hundreds of years, societies have held stormy arguments about what women can and should do. In 1775, the police chief of Paris said women could not possibly stand up to the strain of riding in balloons. He thought women should be protected from the temptations to fly. [1]

Although American women served in various informal capacities from nursing to actual combat since the founding of our country, their formal association began with the military in 1901, as nurses. In 1930, the Army Air Corps said that women pilots were, "Utterly unfeasible," because they were, "too high strung for wartime flying." [2]

The historical World War II perspective of women is the "Rosie the Riveter" stereotype. World War II brought radical changes for women, and although there was progress, very little changed permanently for American women pilots.

American military tradition has never advocated or included women in direct combat. In fact, it has long held the values that a woman in combat is a sacrilege. Over the last fifty years, warfare has changed and so have many social attitudes and institutions.

Although women's records of aviation accomplishments both in civilian life and during the World War II WASP experience should have converted any non-believers, an operations analyst specializing in manpower mobilization stated, "Women's unsuitability for combat is made

---

[1] Ann Hodgeman, Rudy Djabbaroff. *Skystar: History of Women in Aviation.* Atheneum. 1981; p.6

[2] Verne Orr. *Air Force Magazine.* "Finishing the Firsts." Feb. 1985; p.86.

apparent by the fact that they have never engaged in it. Thus... women are unsuited for combat."[3]

This seems to be a Catch-22. Females are unsuited for combat because they have never engaged in it, and the Navy and Air Force would not allow women to fly aircraft that could be engaged in combat. The question arose many times; "Why are women not allowed to fly combat type aircraft?"

In 1948, Congress exempted female Air Force medical personnel from the combat exclusion law so they could serve aboard aircraft flying in combat areas.[4] It appeared at the time that Congress saw the need to "protect" only certain women from combat, but perhaps, not all women. Going back to the Congressional hearings on the Women's Armed Services Integration Act of 1948, (Public Law 625) General Hoyt S. Vandenberg, then Air Force Chief of Staff, testified that the new Air Force wanted women but said emphatically it had no intention of using them as pilots. This policy statement even denied the success of the WASPs who had logged 60 million miles in combat aircraft.[5]

The 1948 Act ensured that women would not have occupational specialties as combatants.[6] Public Law 625 provided a convenient way to exclude women from any position by declaring it a combat or combat related job.

At first the wording, "combat aircraft engaged in combat missions..." was interpreted by the Air Force to mean all pilot jobs, since a pilot should be available for duty in any type of aircraft or any mission, at any time. This automatically excluded women from participating in the primary mission of the Air Force.[7] In the context of the 1990s, the law was sexist, but at the time reflected the prevailing cultural attitude about women's roles and legal rights.

Women have made progress in their roles as equal citizens, and Congress has passed laws with the teeth of enforcement. Title VII, of the 1964 Civil Rights Act is an example. However, Title VII does not apply to military personnel. The combat exclusion legislation continues to discriminate against females in the military. While women now have the potential to earn salaries equal to their male peers, they are still excluded from jobs directly associated with combat. These jobs are the important stepping stones to promotion to the higher ranks. While Congress has responded to the social changes with strong legislation, they have been

---

3
   Jeff Tuten. "The Argument Against Female Combatants," Historical and Contemporary Perspectives: Nancy L. Goldman, ed. Westport Conn. Greenwald Press, 1982; p.239

4
   Congressional Research Service: *Women in the Armed Forces*, CRS-9. Washington D.C. Lib. of Congress; 9/25/85

5
   Maj. General Jeanne Holm, *Women in the Military:* Novato, Calif. Presidio Press. 1982; p.316

6
   Maj. Sandra L. Bateman. *Air Power Journal.* Winter 1988; "The Right Stuff Has No Gender." p.68.

7
   Maj. General Jeanne Holm, *Women in the Military:* Novato, Calif. Presidio Press. 1982; p. 126

deaf to the cries that these same changes should be extended to female pilots in combat. Some in Congress have said as late as 1991, that Americans will not accept its "daughters" coming home in flag-draped coffins. This statement is not true and a contradiction. We did willingly in World War II, Korea, and Viet Nam when hundreds of nurses died in the service of their country, and in Iraq also. Today women respond correctly that they are trained military professionals and in that context not somebody's "daughter." Congress apparently has not chosen to hear this response, at least not from pilots. American military tradition and federal law still does not advocate women in direct combat. The attitude of prohibiting women from combat flying is of course not new, and has been reinforced over the years by some very influential people including Jacqueline Cochran, "I've always assumed," she said, "that we would never put women into combat. If for no other reason than because women are the bearers of children, they should not be in combat. A woman can do almost anything if she works hard enough. But there is something in me that says the battlefield is not the place for women."[8]

Even Charles Lindbergh had ideas on women's place in aviation. "There is no reason why a women should not fly, but they should not be encouraged to enter aviation as an occupation. Their greatest contribution to life can be made in other less material ways. How can a civilization be classified as "high" when its women are moved from home to industry, when the material efficiency of life is considered first and the bearing of children second, if not third."

Naval Aviation

Women first became part of Naval aviation 50 years ago as part of a larger force of "Women Accepted for Voluntary Emergency Service" (WAVES). About 105,000 women served as WAVES and about 20,000 of them were in aviation activities. None were pilots but some were Link Trainer instructors, instrument instructors, and meteorologists. Others worked in air fighter administration.

In 1942, high school junior, Ann O'Hara, wrote to the Navy asking, "Why can't girls win wings like boys? I wanted to be a pilot since I was seven."

In reply a veteran Navy pilot said, "There was no reason women shouldn't fly." He wondered, though, if they could patrol twenty hours at a stretch on instruments over the foggy Atlantic, land on the heaving deck of a carrier, or peal off on a dive bombing attack against a Japanese cruiser. He then admitted, reluctantly, that women could probably do all those things and eventually might.[9]

That was the extent of any real discussion of women flying in the Navy for two decades. There was a brief ray of hope in the sky. On March 25,

---

[8]   Richard C. Barnard. "Where Are They Now: Jackie Cochran" *Air Force Times* Magazine. January 23, 1978. p.18

[9]   Helen Collins. *Naval Aviation News*. "From Plane Captains to Pilots." July 1977.

1966, Ensign Gale Ann Gordon (USNR, became the first woman in the history of the Navy to solo in a Navy training plane, a T-34 Mentor. Ensign Gordon had an M. A. in experimental psychology and was a member of the flight surgeon's class at the Naval Aerospace Institute, Pensacola. Since she would be working with pilots, part of her training as an aviation experimental psychologist was to learn to fly. After Gordon earned her wings the flicker of light faded. [10] She did not become part of an operational unit.

Several years later, the Equal Rights Amendment was sweeping the country. Admiral Elmo Zumwalt, then Chief of Naval Operations, expected ratification of the amendment. He urged then Secretary of the Navy, John W. Warner, to open flight training to women. A voluntary program, he thought, would gain more support than a congressional mandate. Almost 30 years after the Navy pilot had answered Ann O'-Hara, Warner broke the ice that allowed the old salt to see for himself that women could do all those things he had wondered about. Warner announced aviation training for women would begin, and their entrance into flight training marked the first time American women would be trained as full fledged military aviators. The initial group consisted of four women officers on active duty and four women officer candidates. Of this group six went on to become Naval aviators; Ensigns Rosemary B. (Conatser) Mariner, Janey (Skiles) Odea, Joellen (Drag) Oslund, and Anna Marie Fugua, Lts. (jg) Barbara Ann (Allen) Rainey and Judith Ann Neuffer. Three of the six, Odea, Neuffer, and Mariner, had fathers who were combat pilots in World War II.(Fig. 6-1)

The program's objective was to determine the feasibility of using women in future non-combat flying roles, like helicopter and transport squadrons. After the women had completed 18 months of training, received their wings and served six months on flying status, an evaluation of the program would take place to determine its success and future.

Barbara Ann Rainey

On February 22, 1974, Lieutenant (jg) Barbara Ann (Allen) Rainey, a native of Bethesda Md., and a graduate of Whittier College, California, became the first woman to earn her Navy wings. Less than a year after completing flight school at Corpus Christi Texas, Lt. Rainey set another precedent by becoming the first woman to qualify as a Navy jet pilot.

She commented on the experience."Everybody goes through a stage of depression. The hours are long, the work is hard. You sometimes think, 'What's the use? There is so much to learn.. I'll never do it.' I think all students go through that."[11]

Before volunteering for flight training Lt. Rainey was a communications watch officer. In November 1977, Lt. Rainey transferred to the Naval Reserves and flew at Dallas, Texas, for four years. Her two

[10] ibid

[11] Saturday Evening Post. Oct. 1974, p.134

Fig. 6-1 (L-R) Rosemary (Conatser) Mariner, Ens. Janey (Skiles) Odea, Ltjg. Barbara (Allen) Rainey, & Ltjg. Judith Ann Neuffer

daughters were born during this period. In October 1981, she returned to active duty as an instructor in a training squadron.

On July 13, 1982 the career of Lieutenant Commander Rainey ended tragically in a fatal aircraft accident during a training flight. Her student was practicing touch-and-go landings when the plane crashed into the woods during an approach. The student also died. Lt.Cmdr. Rainey the Navy's first designated female aviator also became the Navy's first female pilot killed in an air crash.

Just days before Lt. Cmdr Rainey's fatal crash, student pilot Ensign Cory P. Jones became the Navy's first female Naval Academy graduate to die in an aircraft accident.

Rosemary Mariner

Rosemary Mariner feels it was partly luck that put her in the first group of women to become Naval aviators. It was a combination of being in the right place at the right time, and the result of a burning desire to become a professional aviator. She was an avid reader throughout her childhood and stumbled upon the books of Ernest Gann. Gann so captivated her with his love of flying that she decided this was how one should feel about their life's work. In 1968, in the ninth grade, she told the principal of the Catholic girl's school she attended that she wanted to be an airline pilot.

The Sister didn't bat an eye. At the time there were no women airline pilots. The few women that did manage to make their living in aviation did so predominately through their husbands. The idea of women as airline or military pilots was not realistic. (Fig. 6-2)

Of course such details mean little to a rebellious teenager. Mariner started flying at 15 and washed airplanes to pay for flight time. She earned her private pilot license on her 17th birthday, then entered Purdue University as a geophysics major in 1970. The following spring she was the first women to enter their aviation degree program. Within a year, Mariner had her commercial license with instrument and multi-engine ratings and was a certified flight instructor by age 18.

By this time the women's movement had hit America and the newspapers were beginning to feature 'first female policeman, first female fireman,' etc. type stories. Mariner still wanted to be an airline pilot. In her last year in college she spent time in a Boeing 707 simulator, even though the Vice President of TWA told her, "The American public will never accept women as airline pilots." In the Fall of 1972, Mariner's plans changed. Her mother sent her a newspaper clipping announcing that the Navy was going to train female pilots. It was as if she was waiting for such an announcement. She could hardly believe her eyes. The airlines had yet to hire women, but was the military, that bastion of male supremacy, going to break first? She wasted no time in calling the Navy recruiter. She graduated in the Fall of 1972, and on January 5th, 1973, the Navy accepted her in Flight School.

It is difficult to imagine, if never exposed to it, what massive publicity and sudden celebrity status is like. Mariner enjoyed her special position as the only female in a department of two hundred men at Purdue, but she was not prepared for the hounding of the news media, especially their propensity to quote out of context and sensationalize. All the news services and television networks wanted interviews. Photographers took pictures of one of the women leaning against the tail of an airplane in a miniskirt and boots, and that woman received letters from weirdos for months afterwards. Later the *Air Force Times* was also guilty of sensationalism with the reentry of women to Air Force flight training. The article headlines read, "Dangers to Female Pilots to Be Checked on Planes." The only problem the article identified was the ill-fitting flight suits and boots, originally designed for men of course. "But it was exciting," said Mariner. "There is nothing like watching the Johnny Carson show and hearing yourself mentioned in the monologue," she said. "I was 19 years old, and on top of the world."[12]

While the newspapers were making heroines out of the women their male counterparts were not impressed. The first criticism dealt with that old bug-a-boo, physical standards. The logic was the women were not strong enough to fly airplanes. It wasn't enough that women had been flying since the beginning of aviation, and they had flown every combat

---

[12] Correspondence with author

airplane of World War II. The Navy had forgotten also that the Russians had three squadrons of women flying combat in World War II.

The Navy also seemed totally unaware of history or anything women had accomplished in aviation. One commander said women were physically incapable of flying high performance jets because, "they can't wear G-suits." When Mariner informed him that Jacqueline Cochran had broken the sound barrier in 1953 in an F-86 jet, and that Cochran and France's Jacqueline Auriol had volleyed speed records back and forth in excess of Mach 2, both wearing G-suits, he looked at her as if she were an outright liar.[13] Such criticism and attitudes were the women's first indication that maybe they were not the Navy sweethearts the papers pictured them to be. For the first time for many of them, they were disliked by some people solely because of what they were, women, unwelcomed outsiders, invading a fraternity.

The women received the same training as the men, except for two notable differences. The Navy refused to train women in jet aircraft. The

Fig. 6-2
Rosemary
Mariner

13 Jacqueline Auriol still holds the jet speed record for women

Navy divides its flight training into three "pipelines": propeller, helicopter, and jets. After completing primary training, students apply for the pipeline of their choice. Selection is based on flight grades and needs of the Navy. Not only did Mariner have the grades, but she had finished at the top of her class - by a significant margin. Had she been a man there was no doubt she would have made the jet pipeline. But women could not fly jets because they are combat airplanes, or so they told her. Not knowing at the time how wrong that information was, Mariner and the other women accepted it and entered the propeller pipeline. Like all the women, she was glad to be there, and understood so little about the Navy, it didn't seem all that important at the moment.

The second exception concerned carrier qualification. One thing that becomes readily apparent to any observer of Naval aviation is that Navy pilots, above all, are proud that they fly on and off aircraft carriers. No civilian, Air Force, or airline pilot is their equal without having landed an airplane on the deck of a ship at sea. In carrier pilot ready rooms they brag about their number of "traps" and speak with understated humor of the fear associated with launching off ships at night. In the old fraternity of Naval aviation, landing aboard an aircraft carrier separates the "men" from the "boys."

In 1973 all Navy student pilots were required to carrier qualify before earning their wings. Citing the old "combat law" the Navy ruled that landing a training airplane aboard a training aircraft carrier, the *USS Lexington,* constituted "assignment to a combat vessel." The women accepted the values and standards of their new profession, but also realized that without this important qualification, the men would never fully accept them as equals. The men, many of whom did not want them there in the first place, would say they had been given their wings without earning them. Most importantly, the women wanted to prove themselves, to test their skills, and master the difficult task.

The women finally decided to question the system and submitted a written request for permission to land on a carrier and qualify. They were politely turned down, and left behind on the beach, while the men flew off to the *Lexington.* Besides anger, each woman felt the pain knowing that she could do this thing, but prevented because of an artificial restriction beyond her control. Suddenly the term "discrimination" took on a personal meaning.

Mariner won her Gold Wings in June 1974, and like the other women focused on orders. The men with high grades received coveted "Fleet Orders," to fly airplanes in a "warfare specialty." But all the women, regardless of grades, went to "support" squadrons. These were the same orders that the men who finished at the bottom of the class were given. But at the time the women were more concerned with obtaining the best possible set of support orders than fighting what seemed a no-win battle. With little remorse the women took orders to fly transports in Spain, helicopters in San Diego, or hunt hurricanes in the Caribbean.

Carrier Qualification

Becoming a "carrier qualified" Naval aviator for women was an evolutionary process. Legal action, general dissatisfaction, the Women's

Movement and the Affirmative Action program turned the polite "no" to an "OK, let's see if they can do it." As a result, Particia Denkler became the first woman to carrier qualify in a jet. (Fig. 6-3) Mariner became the first woman to fly a tactical jet aircraft, the A-4 "Skyhawk" in 1975. Women became carrier qualified, in props only, in 1979, but began to earn their wings in the full jet pipeline, including tactics and carrier qualification, in 1982. Today carrier qualification is a routine event for women aviators. Although most women are still in transports and utility aircraft, women like Mariner, are flying front line tactical airplanes like the A-7. Unlike the Air Force, which excludes its women from tactical aircraft, there is nothing to preclude Navy women from flying F-14s, or even F-18s, in shore based commands, which now some do.

Navy women routinely fly the mail to aircraft carriers and deliver food and weapons to ships all over the world. Many marry fellow aviators, and successfully combine marriage, career, and motherhood. Beside the many who fly passengers and cargo, other women drop bombs, and simulate Soviet air tactics in jets against fleet pilots. They dogfight, test new missile systems and serve as flight instructors. Women flight students are now a common sight in Pensacola.

There are many contrasts that testify to the changes in the last twenty years. Today women live and work on ships that until a few years ago they could not board unless in civilian clothes, on leave, and accompanied by a male officer. Behind all these steps forward lay a series of hard fought battles. The old timers remember the frustrations, and a few ruined careers. Behind each of these "firsts." Mariner says, "We are keenly aware that so much more was, and is, possible." For example, the Navy could train over one hundred women pilots per year without placing them in combat squadrons. Before the Iraqi War, women helicopter pilots deployed aboard ship to the Mediterranean Sea, but not the Indian Ocean. Before the "Women at Sea" program made the news, several women officers won a lawsuit to earn the right to go to sea. Even after it was clearly legal for women to land aboard carriers, the first woman to do so had to transfer from the West Coast to the East Coast because a certain admiral wouldn't allow women aboard "his" ships, and the jet pipeline did not open to women until a near-fatal accident in 1981.

Rosemary Mariner was asked why she wanted to participate in combat. Her answer was, "My reasons are the same as those that have always attracted men to Naval Air. It is because I have experienced the satisfaction of the first step - winning Gold Wings - and I want to continue to succeed at what is the most demanding form of aviation. I want to become a full professional in my chosen vocation.[14]

The biggest roadblock facing the women may be dissolving. Until recently there was no fundamental change in the career path of female aviators. Women for the most part were still assigned to the secondary and tertiary "support" groups; groups that are not career enhancing for

14 Helen Collins. *Naval Aviation News*. "From Plane Captains to Pilots. July 1977.

Fig.6-3 Patricia Denkler First woman to carrier qualify in a jet aircraft.

men. In 1986, Mariner was selected for aviation squadron command. In 1987 she reported as the Executive Officer of VAQ-34 (an electronic warfare squadron). There were nine female aviators in that squadron alone, and was the basis for the women pilots in the TV series "Super-Carrier." Eighteen months later, Commander Mariner assumed command of the squadron - another barrier had fallen.

In 1990 there were 12,477 Navy pilots. There are 225 women pilots none of whom have reached the rank of captain. To be fair, it takes about 20 years to earn those four stripes, and women haven't been officers in naval aviation that long. Theoretically there is a chance under the existing combat exclusion laws for a woman to command a carrier as a captain. There is one aircraft carrier that never sees combat, the *U.S.S. Lexington*. If this ever happens, it will probably seem to many women as a concession, and not an acknowledgement of their true value. Only time will tell.

In general, service policies are more restrictive than the laws themselves. While it is possible for women to attain equal opportunity, these laws and policies prevent them from acquiring the necessary experience to compete for the high level operational commands that are the pinnacle of a military career. Today it is impossible for a woman to even remotely qualify for the top job in the Navy, Chief of Naval Operations. Women have a better shot at becoming Secretary of the Navy. According to Mariner, "No matter her ability, the determining factor in a military woman's career is still her sex. There can be no such thing as equal opportunity until the combat laws are removed."

In order for this to happen the military itself must initiate action, and aggressively petition Congress for repeal. This could be done in the same manner military leaders initiate legislation for changes in personnel laws, actively push for increases in compensation, and publicly support controversial weapons programs.

The basic issue is still women in combat. Just as men accept the risk of combat as a necessary one in the military profession, so do many women. There should be little doubt that women can fight, they have been proving this throughout our history. The issue again is whether women should fight.

Amelia Earhart discussed this in 1935. "Drafting women for the real work of war, not the sideline jobs where you can wear giddy uniforms and not get dirtied up, would cause much discussion of such a draft. We would hear chivalry crying out that females are much too frail for the inhuman cruelties and hardships of what is, after all, a man's game. Nonsense! Such a concept of chivalry is hypocritical. Combat service in the air, transport jobs in advance positions, and even other less brilliant arenas of activity in the theater of war, are the last remaining strongholds of men. I suspect that men might rather vacate the arena all together than share it with women." Women flew transports, helicopters and tankers in the forward areas of the Iraq war theater - a step closer.

"I have spent all my adult life in the macho world of military fighter pilots," said Mariner. "Women who have spent some time in non-traditional, male dominated fields probably understand how different that male world is.

"Sometimes I feel like an immigrant, reflecting upon my life in the promised land. I have learned to speak the language well (including the four letter words), have experienced success beyond my wildest dreams, and totally assimilated the values of my new country. But I have also learned the standards of this culture and understand that I, and those of my generation, will always be foreigners. We are the ghetto dwellers, excluded from the mainstream, and told we should be thankful for all that has been "given" to us. For no reason other than an accident of birth, we are subjected to many restrictions fully sanctioned by the law. I now have some understanding of what it is like to be called a "Wop," a "Jew Boy," or "Nigger." Our only hope for total integration lies in future generations. But like that same immigrant, I do not regret my choices. In my own career I have been very fortunate. I have seen many male officers support women, some even risking their careers to do so." [15]

Because of that support, Mariner has a whole series of "firsts" behind her name, first military woman to fly a tactical jet, first woman to fly the front line A-7E, and so on. She has also been the subject of much publicity and controversy. The A-7 is a combat airplane, normally based aboard carriers and used extensively in Viet Nam and Iraq. Because she flew it in a research and development role she did not violate the law. For five

[15] correspondence w/author

years she dropped bombs, fired guns, and tested new weapons systems in the Mojave Desert.

For the most part, she has enjoyed working with the vast majority of outstanding men who are secure enough in themselves not to allow whatever doubts they may have about military women to interfere with the job. "I have nothing but the highest respect for the men who fly off aircraft carriers on dark nights, in pitching seas," said Mariner. "I feel honored to have shared the camaraderie of their ready rooms.

"And I have seen the negative side of flying. I have been so scared, all alone at night in nasty weather, running out of fuel in a malfunctioning airplane, that I swore if I ever got down safely, I would never do this again! But six hours later I went up again."

Mariner has sifted through the wreckage of many airplanes as an investigator, trying to find an answer to why a friend, so alive yesterday, was splattered across the desert floor. "I have watched my wingman crash and die and known the feeling of total helplessness, for I could do nothing to save him. One can get killed in this business, male or female, without ever being shot at by an enemy.

"But I have also known the total exhilaration that comes from flying an airplane at 500 knots through the weeds, and putting one's bombs right on target. I have seen the great beauty of this magnificent country from 30,000 ft., and wondered out loud, "I get paid for this!"[16]

Although the Navy has roughly 45,000 women on active duty (about 7,000 officers) this is roughly 9% of the total force and only half the number of WAVES in 1944. Female naval aviators are a very small presence just over 200. The Air Force opened its program four years after the Navy, and has about 400 women pilots. But none of them are flying weapons or tactical jet aircraft. Navy women, however, do all these things as part of getting their wings in the jet pipeline. In retrospect, there has been progress. In 1972, 99 percent of the Navy's women officers served in just two types of squadrons. In 1987, women officers were found in all but two types of squadrons: submarine and special warfare.

Mariner also cautions, "The military is not a training ground for future airline pilots. If you want to be a Navy pilot, you better be willing to go to sea, and if necessary, go into combat; male and female." Her advice to young women thinking of an aviation career, "Go for it! I can't think of a better career."

Judith Ann Neuffer

At one time, only men volunteered for one particularly tough Navy job, flying through a hurricane. That was before women had flying status.

Lt. Judith Ann Neuffer was in the first class of women to through Navy flight training, and she was also the first female pilot with the Navy's Hurricane Hunter Squadron at Jacksonville, Fla. She met the challenge

---

[16] ibid

of peer acceptance and flying in, out, and over, the world's most destructive storm with professionalism and dignity. (Fig. 6-4)

Born in Wooster, Ohio, Neuffer grew up around aviation. Her father, an Army Air Force pilot in WW II, started giving her flying instruction at 15 and she soloed in a Piper Cub at age 16. Lt. Neuffer came to "the toughest noncombat flying job in the world" by way of Ohio State University. She asked for orders to the Hurricane Hunters because she wanted the challenge. "I never had time to think about how I felt in a hurricane. When we were in a storm I was very busy. You are totally committed and you're concentrating on what you are doing. Your only purpose is to get into and out of the storm safely. I was never conscious of anything other than using my training to fly the aircraft properly. I'm sure subconsciously I was feeling fear because I was very tense. But for the most part I spent my time concentrating on what I was doing." Lt. Neuffer brought the four-engine P-3 Orion aircraft through her first storm like most pilots, with slightly frazzled nerves. "All I could see was a white wall of clouds," she said, "It became a routine experience for us." The routine for her was wind speeds of more than 150 miles per hour, blinding rain, downdrafts and turbulent eddies.

When Neuffer enlisted, pilot training was not open to women. Neuffer earned her commission as a computer programmer, and while in that job the Navy opened the flight program to women.

Lt. Neuffer had to prove her ability to fly with the elite Hurricane Hunters but she also had problems outside the squadron convincing people that she was a Navy pilot. "We were new in those days, and nobody thought I was a pilot. When I stepped down from a plane after a flight, they'd look at my green uniform and black boots and thought I was a nurse along for the ride."

Neuffer had peer support too. "I could not have asked for better support than I received over the years. The men are outstanding. It's hard to say enough good about them. That is not to say there wasn't resentment along the way because that was inevitable. I was doing something different, a change from the way things had been. It's harder for some people to accept change. When people see I'm not out to prove anything, or to pose a threat to them, I'm accepted."

Neuffer made the point that a woman does not have to sacrifice her femininity along the way. "I don't want to be one of the guys, just a member of the team. There is a difference and I think I've succeeded. I'm a woman and I'm very glad to be a woman. After the men have flown with me and realize I'm just like any other pilot, they simply treat me like a member of the crew, with no special considerations."

Neuffer thought for a moment. "Well, maybe one. The language on my crew is not as colorful as it is in others. I respect and appreciate that, but I've never asked for it."

Nueffer also discovered what many military women have found. "I had very little time to make contacts outside the squadron, so my social life was nothing out of the ordinary. I know a lot of women think, "What odds - one woman and all those men!" Even though there are some advantages there is also a lot of hard work. I get grubby in my flight suit, and at night I spend most of my time studying. It's not always glamorous.

Fig. 6-4 Judith Neuffer

"There were times where I would like to have a woman to go to lunch with or to talk woman things with. I just can't go to the wardroom and sit down to talk about woman things with the guys. So there are advantages and disadvantages."

Neuffer feels she is an average person. "I'm not gifted or in any way superior to anyone else. Motivation is the key. It's a lot of hard work. I've never worked harder in my life, but it's been worth it. I've seen the world, from one pole to the other. I've certainly had opportunities that many women haven't had. I've been lucky."

Colleen Nevius

A decade after the Navy admitted women to flight training, another barrier fell. In June 1983, Lieutenant Colleen Nevius, became the first woman to graduate from the U.S. Naval Test Pilot School (TPS) Nevius isn't impressed with the label, "The Navy's first woman test pilot,"and feels she doesn't deserve special attention. "The people who deserve recognition are those who laid it on the line and selected me," she remarked. "They gave me the chance to stand or fall on my own."

Colleen Nevius decided to become a Naval Aviator in 1973, the year the Navy opened its flight program to women. "I didn't want to pursue anything else career-wise," said Nevius, who was an 18 year-old high school honor student at the time. "Besides, I knew I was physically and mentally capable of flying. It just seemed like the right thing to do."

Nevius became one of the first women to receive a full Navy ROTC college scholarship. The scholarship paid for her tuition, books and fees, plus a $100 per month in return for four years of obligatory service.

In 1977, she graduated from Purdue University with a degree in management and the following year she entered the Aviation Officer Candidate School, in Pensacola, Florida. During her primary and advanced flight training in Corpus Christi, Texas, Ens. Nevius flew many of the same T-28 Trojans her father had flown while he was a flight instructor there in 1955. They later compared log books and discovered that she soloed the same T-28 that her dad had flown 23 years earlier on the day after Colleen was born.

Following flight training, Nevius went into the helicopter pipeline. "After having flown a small prop plane and a jet, I climbed into a Huey and I fell in love with them on my first flight," she said. "They're also more versatile and they are fun." In addition, the realist in me figured that helicopters might be the first aircraft in which women could join an operational warfare community like helicopter combat support."

After successfully completing helicopter training, Nevius received her aviator wings in February 1979. Following flight training, the Navy assigned Nevius to a Helicopter Combat Support Squadron in Norfolk, Va., where she worked as airframes branch officer and helicopter aircraft commander for the squadron. While there, she and another female pilot became the first women to fly vertical replenishment helicopters to ships at sea. However, Congress changed the code and permitted women to deploy on ships temporarily.

During her tour, Nevius flew medium and heavy lift helicopters (H-46 and CH-53E) and made more than 500 shipboard landings on more than a dozen vessels. "Aboard ship is the best place to fly," she said. "It is the most challenging and the most rewarding. We are out there doing the same thing in peacetime that we would be doing in wartime." Nevius decided to try for the U.S. Naval Test Pilot School, (TPS) which has a grueling eleven month curriculum aimed at turning some of the most exceptional Naval Aviators into test pilots. (Fig. 6-5)

"Test Pilot School sounded like a challenging tour." she said. "As perhaps in any selection process, I was not at all confident about the outcome," she added. "I hoped my record and experiences would speak for me." They did. TPS, which had been a bastion for training only male Naval Aviators since its inception in the mid-1940s, chose Nevius in 1982. Her acceptance enabled her to attend a school that has been influential in the careers of distinguished men like John Glenn, Alan Shepard, and Scott Carpenter, to name a few.

Today's Navy Test Pilot School is different from the silver screen view. It is no longer for the Chuck Yeager, "yank and bank" types. Now they emphasize mild-mannered engineer types who are more apt to think problems through instead of using brute force and trying at the same time to kill themselves. "TPS was 150-percent demanding," said Nevius. "It required a lot of time, perseverance, dedication, talent, and a desire to survive. The school got more out of me than I thought I had, but it was most satisfying to complete."

"You can get burned out easily because the pace is fast and you never feel you've done enough," she said. "You spend about a year at the Test Pilot School, six months trying to learn the ropes, then 18 months as a project pilot." As a test pilot, Nevius' work week averaged between 40 to

100 hours, depending on the project load. She added that each day is different, involving flying reports, phone and personal liaison, travel, rest planning, review of test results, and research.

"My fellow officers treated me as a fellow officer and pilot. When I did dumb things, I got my share {of criticism}. When I did okay, life went on." Nevius also became the first female member of the Society of Experimental Test Pilots.

She added that the Navy should try to broaden the opportunities for women. "I believe that all individuals should have the choice to pursue their talents and desires without limits," said Nevius. "And I look forward to the day when there aren't any restrictions on what women can and cannot do in the Navy."

"There will always be some people who react first according to gender," she added. "But, with luck, they'll learn that there are more important differences to consider in the work place."

Today Lt. Cmdr. Colleen Nevius is flying C-12s (KingAir 200s) for the Naval Reserve, while working on her Master's Degree in Computer & Systems Engineering at the University of Houston, and doing part-time consulting in the Johnson Space Center area. In 1991, Colleen Nevius became executive director of the Whirly-Girls.

Fig. 6-5
Colleen Nevius

Fig. 6-6 Janey
(James) Odea

Fig. 6-7 Joellen
(Drag) Oslund

U.S. Air Force

The long-held historical perspective of June Allyson watching through tear filled eyes as her man flies off to war is just that, history. Women entered Air Force pilot training in 1976 and navigator training in 1977, but it was not until 1984, when women could fly the KC-10 tanker. About 12.5 percent of all Air Force personnel are women although that includes non-flying jobs. Women are equally eligible as men for pilot training except for that specifically designed for combat. There are more than 33,000 officers rated as pilots and navigators and approximately 400 are women pilots and navigators. [17] The Strategic Air Command has about 100 female pilots and about the same number of female navigators. The statutes still prohibit women crew members in bombers so these women fly tanker aircraft.

Connie Engel

When the Air Force opened pilot training to women, Connie Engel became the first woman to win her wings. She began her Air Force career as a nurse and originally had no intention of going into aviation. She admits that there may have been some subtle influence working; Connie's father was an Air Force pilot. Her husband Rich, also an Air Force pilot, worked on the project coordinating the first class of women pilots. The two became involved in the planning, and one thing led to another. Connie did not feel she was growing in her career, and her husband encouraged her to apply for flight school.

Her decision was not easy. She would be charting new territory. Connie was accepted to flight school at Williams Air Force Base, in Texas, and things were looking up for the Engels. Then came a test of their love. Rich received his appointment to test pilot school, at Edwards Air Force Base, in California. It would mean a one year separation between them with a few weekends together, schedules permitting. (Fig. 6-8)

Once Connie entered flight training she had no major difficulties just some frustrations and adjustments. For awhile her physical appearance did suffer. "Wearing men's flight gear (Boots, suits, helmets,) was difficult on the feminine identity," she said. "We weren't allowed any makeup (it interfered with the oxygen mask) and our hair constantly looked like we'd just taken off a swimming cap. We were in a field that required physical stamina, long hours of flying and studying, and lots of running, parachute training, parasailing, etc. This physical world was new to me. I was always the cheerleader, not the football player, and it hurt lots of the time! But, I learned to be tough."

Being the first female pilot brought on a lot of criticism (usually in the form of gossip) which had to be handled so it was not detrimental to Connie's performance or everyday life. "I learned to turn to the Lord for

---

[17] Air Force Fact Sheet 87-45

my strength and identity, and just let comments roll off. For the most part they were not true and spoken by thoughtless bystanders."

There was a lot of pressure too. Many people were interested in her progress, achievements and problems. The commanding officer of the Air Training Command told her, "Your success will have a direct and lasting impact on the Air Force."[18]

Kathy La Sauce, one of Connie Engel's classmates later commented on the pressure. "When you step out of the airplane and take your helmet off, the transient maintenance guy almost falls over backwards. That's the sort of thing that keeps us going. There was a great deal of pressure to be good, to succeed, or the door would be shut for other women." [19]

Connie was the first in her class to solo. In a class of ten outstanding women and thirty-six outstanding men she was the class leader. (Fig. 6-9) On graduation day she received the Officer Training Award for, "the student who has exhibited high qualities of military bearing and leadership." She received one of the three "Distinguished Graduate" awards,

Fig. 6-8 Connie Engel

18 Correspondence with author

19 Terry Arnold "Baptizing the New Breed." *Airman* Oct. 21, 1977 p2-7

and the most coveted of all, the Commander's Trophy; for overall excellence in flying, academics, and leadership. The most significant award of all was the one her husband Rich presented to her. Rich pinned on her silver wings. Engraved on the back she read, "That we may slip the surly bonds, and touch the face of God, together."

Many pilots make personal sacrifices along the way, but for a married pilot the most obvious sacrifice is that of the spouse. There were long hours of sharing the workload, sharing the "fish bowl" that the woman pilot occupies, and being rejected for the "famous wife." "I must credit Rich with so much. He was an integral part of my success. He was encouraging, helpful, excited, and accepting of the time and effort required to be in this career field," says Connie. "Flight training was twelve months of our lives, 14 hours a day, heavy studying and overnight flights."

To be a career Air Force pilot takes a strong committed individual. It is exciting and fulfilling, but as Connie can testify, the price is high. "The women in America need to have no illusion about pursuing a long career as a professional military pilot. It will mean a sacrifice in the out years - either their career (as mine) or their family.

"The children in America are for the most part growing up without their mothers at home. A new generation of "day care children" who lack the security that God designed them to get from the care of a full-time mother. If the woman desires to have a husband and family the choice must be made somewhere during her career."[20]

Connie left the Air Force with the rank of Major. "In making the decision to leave the active Air Force after 13 years, I gave up a career of being a woman on the road to leadership of the Air Force. Instead, I chose, as God instructed me to choose, a career that attracts even more negative comments than the woman pilot, that of the full-time homemaker."

A second issue facing a female Air Force pilot is that of women in combat. Military women are at the forefront of the congressional battle. Connie feels that in time, Congress will include women in the frontline warfare of this country but doesn't agree it's right. "While many of us would agree to go to war each must make their decision before going into this field. While trained to fly a fighter aircraft, I personally feel it is a step backward for America to include their women in this protective role which God designed for our men."

On a lighter note, Connie summed up her attraction to flying. "I love the excitement of turning upside down in God's creation, being alone and in control of such a powerful machine. I enjoy the fruit of seeing a professional in action to accomplish a job that I have been trained to do. I like the immediate feedback of seeing parameters met with perfection like in formation flying. It's just about the most exciting thing I have ever done, and it stimulates my emotions in every realm, excitement, tears, and fear. It is an accomplishment."

[20] Correspondence with author

Fig. 6-9 Connie Engel's class (L-R) Susan Rogers, Mary Donahue, Cris Schott, Vicki Crawford, Connie Engel, Kathy LaSauce, Carol Schearer, Sandy Scott, Mary Livingston & Kathy Rambo

After Connie earned her wings, acceptance as a pilot was not automatic. "There is a time of demonstrating, proving if you will," said Connie, "your ability to be trusted out there with another aviator who depends on you to perform, for their safety and accomplishment of the task. Women, the unknown entity, have to go through this, too. However, as is the case in numerous other professions, women have to perform "better" than the male to earn the same "average" to "above average" rating. The really average woman is usually considered a poor pilot in comparison to the average male. The exceptional woman pilot has to work twice as hard as the exceptional male pilot. Acceptance is earned on an individual basis with individuals. Many of my peers and superiors disliked women pilots, but liked and respected me. They were won one at a time, and one by one, became firm and loyal supporters."

Connie Engel is satisfied in the role God designed her to fill. "I am not ashamed to serve Him, meet the needs of my husband, and to raise my children in the nurture and admonition of the Lord. I do not feel being a professional military aviator is exclusive of these priorities, but they all have to fit into their proper order. Because we serve Him, we are able to fit all things into their proper order."[21]

Steps Forward

On March 23, 1978, Captain Sandra M. Scott became the first female pilot to perform alert duty in SAC. Captain Scott was in Connie Engel's class and assigned to the 904th Air Refueling Squadron, a KC-135 unit at Mather Air Force Base, California.

"Fair Force One" - SAC's first all-female crew made a historic flight on June 10, 1982. The crew from the 924th Air Refueling Squadron at Castle Air Force Base, California, performed a five-hour training mission that included the mid-air refueling of a B-52. Nicknamed "Fair Force One" by operations specialists, the crew was commanded by Captain Kelly S.C. Hamilton, SAC's only female aircraft commander at the time. The other crew members were: First Lieutenant Diane Oswald, navigator; and Sergeant Jackie Hale, boom operator.

Two other all-female teams supported the mission. The personnel who scheduled the mission were Technical Sergeant Lorrie Roberts and Staff Sergeant Laurie Moffatt, both assigned to the 93rd Bomb Wing's training operations branch. "Fair Force One" was not a permanent crew, but the all-female ground crew who supported the flight had worked together for some time. Its three members, who made up the first all-female ground crew at Castle Air Force Base, included crew chief Staff Sergeant Judy Ricks, assisted by Airman First Class Gwendolyn Hoyle and Airman First Class Terese Sessums. While pleased to have taken part in the special mission, Captain Hamilton pointed out that permanent all-female aircrews would not become policy since doing so "would defeat the concept of further integrating women with men in the Air Force."

Granada

Supposedly, the Air Force is totally integrated except for combat positions, but female pilots flew troop, and cargo-carrying missions to Grenada during the initial phases of the 1983 invasion. They landed on the Caribbean island aboard C-141 aircraft when US paratroopers were still fighting Cuban troops. A male pilot who flew to Granada said, "The significant thing is that they went in, did the job alongside us, came out, and nobody made a huge fuss about it. Nobody made a special effort to include them and nobody thought for a moment about excluding women." To have excluded an aircraft from the mission simply because there was a woman on board would have lessened the response and reduced the mission's effectiveness."

Another example of how reality has overcome an outdated law is the counter-terrorist attack the Air Force flew against Libya in 1986. Seven women, six officers and one enlisted, served in the raid. One of the

[21] ibid

women was a backup pilot on a KC-135 tanker, and four served as copilots, three on KC-10s and one on a KC-135. Secretary of the Air Force Edward Aldridge, Jr., stated, "Women flew on these aircrews as a natural evolutionary growth of the contribution of women members to the Air Force."

Another frequently heard argument supporting female combat exclusion is that females in fighter units will impede the "bonding cohesiveness" of the unit, which will degrade mission effectiveness. The fact that this has not happened in squadrons where females have flown for the past 10 years does little to discourage the myth. Many female pilots have said once the men realized women can fly as well as the men, the discrimination evaporated. Astronauts have to work together and under more extreme conditions than the average pilot, yet they have exhibited no problems "bonding" with their female crew-members.

Senator William Proxmire, of Wisconsin said, "Current assignment policies do not really protect women from combat and are a waste of talent. The range and effectiveness of modern weapons make it impossible to isolate female soldiers from the danger of combat. The support jobs they are allowed to take are often as dangerous as the front-line jobs they are now prevented from taking, as witnessed in Iraq.

Another Air Force officer said, "A woman cannot fly a fighter or a bomber, but she can fly a tanker to refuel them. Suppose you are an enemy fighter pilot and you have one missile. What do you shoot down? If you get the tanker, with the female pilot the bombers will fall too!"

Commander Rosemary Mariner summed it up best, "I will say this. Based on my experience, I have no doubt that women (pilots) will fight, and die, in our next major war. From a pragmatic viewpoint based on the demographics of the next decade alone, we cannot afford the luxury of wasting one half the population in what may well be a do-or-die war. Especially if women and children die in their own backyards first."

U.S. Coast Guard

A 1790 Act of Congress created the Coast Guard, to stem the tide of smuggling by enforcing the import tariffs. Over the years, its role has shifted but generally it is responsible for protecting the territorial waters of the United States and its possessions. Today its primary mission is to act as a coastal sentry. Coast Guard aviators participate in search and rescue, both in and over oceans and inland, law enforcement of fisheries laws, and apprehending drug smugglers. Sometimes they carry troops (Bahamian for example) to capture drug smugglers.

As the Coast Guard's role expands so have career opportunities. The role of the aviator is an essential part of these missions, but only recently have women joined the ranks of the Coast Guard aviator. The reason for this is not official policy excluding them, bureaucratic or command decisions. It goes much deeper.

The Coast Guard, unlike the Navy, does not consider aviation its principle arm. "Aviation is what's left if you can't command a ship," said one pilot. The commandants, she points out, have almost always been ship drivers. Only two have been aviators. The opposite is true in the Navy.

Coast Guard aviation is also underplayed. They advertise they want qualified people, men and women. Opinion is mixed on how to get them. Some aviators say go out and recruit more women. Others say, "Here are the qualifications and sex is not one of them. We don't care if you're male or female; as long as you can do it, you get the job." The Coast Guard recruiting budget is also small, some say too small to attract good people. It is word of mouth recruiting, some say. Fifty percent of Coast Guard aviators come from other services. Less than half the aviators come from the Academy and the Academy doesn't stress aviation.

On March 4, 1977, Ensign Janna Lambine earned her Gold Wings and became the Coast Guard's first female aviator. Lambine remained the only woman for awhile since the Coast Guard had shown only a limited commitment to training women aviators. The Coast Guard was sensitive to the public pressures that demanded at least the appearance of integration.Ensign Lambine experienced the isolation that was typical of the first women entering the "fraternity."

Claudia Wells

Lt. Claudia Wells attended OCS in Yorktown, Virginia, in 1979. Her first tour of duty was aboard the Coast Guard Cutter *Munro and Mellon* stationed in Seattle. Her second tour was as an OCS instructor. In 1983 she decided to stay in an operational billet (without going back to sea) so she applied to flight school. Wells says most men and women have been supportive and she had no difficulties. "I was the first female aviator at San Diego and welcomed with open arms."

Wells entered the service with the idea of making it a career and nothing has occurred to change her mind, she said. "My miliary career disrupted my social life because of the many moves but overall the sacrifices have been worth it. I have been accepted by my colleagues. I did have to pay dues on the ship to win over the Warrant officers and the Chiefs but all Ensigns do. Once the Warrant Officers accepted me, they became my strongest supporters.

"I feel very fortunate. Part of this is because of my age (Wells entered the Coast Guard at the age of 26 and because of her banking career before her enlistment had the opportunity to supervise men and women.) I feel that helped me. I hope more women will enter aviation, it can be a very rewarding career."

Alda L. Siebrands

Lt. Alda Siebrands has about 18 years flying experience. "I was in the Army as a communications officer and one of my fellow platoon leaders had gone through ROTC flight training. He was awaiting his slot in flight school, and that started me thinking about aviation. I took the physical and aptitude tests and within six months I was in flight school."

Flight school provided the same challenge and difficulties for her as it did for her twenty male classmates. Her first duty assignment was in Korea. There she became the stereotyped administration officer under a Major who had never worked with a woman pilot before. "It was difficult for him to accept that I wanted to fly like any other new pilot so

I had to sneak off to fly with some of the other guys to get on flights. That worked itself out within six months and I had no problems after that."

Siebrands ,who has more than 4,000 flight hours ,discovered that once military men found she could do her job, they became her supporters. The women who are non-aviators look up to her as an aviator. "Civilians are always curious about my job as a helicopter pilot, and women are curious about how I work with the men."

"There are just a few dozen women in Coast Guard aviation, and a person's assumption is I must be the copilot or crewman. When I flew in the Gulf of Mexico at a civilian job I got some interesting reactions from the oil workers when they realized I was the pilot that would get them off the rig and back to shore. Many of the workers became familiar with whom I was after having flown several trips with them. They became possessive of me as their pilot. They had their own notoriety because they flew with the female pilot. In Korea, it was difficult for the Koreans to accept that I was a woman because obviously a woman can't fly.

Siebrands, also a Whirly-Girl said, "Many women do encounter some initial coolness from the men. But you do your job to earn their respect. Some men are very susceptible to letting a woman get by with things because their expectations for you are low." It will be awhile before women are fully accepted," Siebrands thinks. "I still run into men who have not had their first experience working with a woman pilot." Siebrands was at one station for the three years and never ran into another woman pilot. What she considers more shocking was that there were no female crew members, enlisted, either. "We had 23 pilots and 40 crewmen and I was the only woman. As long as that situation exists, "acceptance" will be difficult for the men. More flight time seems to be the equalizer. I am very proud to be an aviator. It is a great way to make a living."

Laura Huffsteller Guth

Lt. Laura Guth received her wings in October 1980. Guth doesn't remember when the desire to fly first began but she does remember wanting to fly for a long time. "It never occurred to me, as a girl, that I couldn't do what I wanted to."

Guth says reactions of men and women to her as a pilot have been funny, serious, and sexist. They have run the whole gamut. "My favorite story is a recurring reaction I get, usually from older ladies. I am only 5' 4", but fly large (22,000 lb.) helicopters. At airshows, I usually get at least one person who finds out that I am the pilot, looks at me, looks at the aircraft, and says "You mean a little bitty thing like you flies that great big helicopter?"

Guth too has paid her dues many times over. "Every time you went to a new command you had to prove yourself all over again. It's getting better now. Nobody was willing to accept that you made it through flight school, you earned your wings so you must be able to fly the aircraft.

"Every time I made a different qualification, like moving up from copilot to Aircraft Commander, it started all over again. It was, "well you were OK to fly as copilot but you have to prove yourself as Aircraft

Commander. I got more support from my CO and department head than my peers. To a large degree they felt threatened.

"I left the Navy and went to the Coast Guard in a large part because of the treatment in the Navy. A lot of the prejudice is an institutional attitude. For example when the Navy said they were going to accept women in flight school some accepted the decision, some did not. It was condoned not to accept it. It was OK if the admiral didn't want women flying. The Coast Guard has taken the attitude women are going to fly and if you don't like it you don't belong in the Coast Guard."

Guth does not what to fly anywhere else. "I wouldn't like to fly for an airline. What I'm doing is exciting. Flying an airline is like being a big bus driver with no excitement, and the chauvinistic attitudes still exist. I was out doing instrument approaches one night," said Guth. "Our approach control heard one of the women talking from an airliner. The controller was asking if she could get a little more speed. She replied, 'we'll try and give it all we can but we must have a dog of an aircraft tonight.' Some guy came back on the radio and said, 'All you need is a better driver!' I thought we were getting beyond that. I was shocked."

Colleen Andersen

By 1988 one would think enough medical data on women pilots was available to make sound and intelligent decisions on flying and its effect on the physiology of women. There was the WASP experience during World War II where they were tested in high altitude chambers and subjected to other physical tests. There were two false starts in the space program where they again were subjected to a multitude of physical and psychological tests. In all results they did as well as men and sometimes better.

Lt. Colleen Andersen pilots a Falcon jet for the Coast Guard. She too had to demonstrate her ability and found the struggle for acceptance of female pilots does not always end in the cockpit. In some areas of the military there is still a lot to learn about the differences between men and women. (Fig. 6-10)

The most obvious area to women officers is that the military establishment's attitude is 30 years behind the times. One area in particular is the uniform for females in the Coast Guard. In 1988, Women officers were still wearing polyester pants that zipper in the rear instead of the front. They have no pockets or belt loops to help keep their shirts tucked in and their shirts have no pockets to help line up ribbons and name tags.

The presence of women in a group of officers is sometimes made obvious to everyone. When a male speaker is addressing the group he will usually hesitate during his welcoming statements, i.e. 'Good morning gentlemen ... and ladies. Sorry about that.' "Women like to be acknowledged" said Andersen, "but not to the point where the speaker feels obligated to single out or include females after he has forgotten."

The Coast Guard knew that women officers would eventually be stationed at Air Station San Diego, and yet it took almost four years after the first woman pilot arrived before a woman's locker room with shower was built.

In flight school, Andersen was attending a lecture and the instructor stated that because a woman was present in his class he would be unable to tell dirty jokes. She told him to, "Go right ahead," and that only embarrassed him further. Said Colleen, "In the end, I can only be humored by those who make themselves feel uncomfortable around women. In the field of aviation I regard myself as a pilot first, and a woman second, because I know that I can do the same job just as well or even better than my peers."

Colleen was the first pregnant pilot in the Coast Guard and every step of the process was blown out of proportion. Under most medical circumstances the Coast Guard follows U.S. Navy policies. In this case, however, Colleen was grounded at four months, instead of six months because all avenues of research had not been explored. Some individuals refused to look at the facts. The doctors thought that Falcon jet pilots wore "G suits." The problem was they didn't ask. Had they made a simple inquiry, they would have discovered that a Falcon jet is very much like an airliner.

But to be fair, not enough study had been done on pregnancy and airline travel or piloting. No one really knows the long range effects of 8,000 foot cabin altitudes or high altitude radiation on the fetus. The only problem Colleen might have had while flying beyond four months was bladder control. Her longest flights were four hour patrols and there are no relief tubes in the Falcon.

Even the FAA did not have an official policy. When Andersen went for her FAA First Class Medical Certificate during her pregnancy, the

Fig. 6-10
Colleen
Andersen

medical examiner passed her premature grounding experience on to an airline that was trying to establish a policy for their women pilots. Andersen said, "It's time the military accepts the fact that women pilots are here to stay, and there are certain areas that have to be recognized and addressed."

What bothered Andersen the most was that she was only granted the mandatory 30 days of maternity leave. "I had purposely not taken any leave during my pregnancy so I could take an additional 30 days. My intentions were not relayed to the command as I had thought. I was called the third week after the birth and told to report back to work the next week. I was totally unprepared to stop breastfeeding, lose weight and find a permanent babysitter, all within a week. The reason I found out later was the commanding officer felt that the other pilots had carried the load long enough. If the military is going to accept women they must understand the importance of breastfeeding and establishing the mother-child bond." (Enlisted women are allowed to request early-outs for these reasons but women officers are not even given an option.)

In 1939, the CAA had temporarily banned pregnant women from piloting. Betty Gilles, then president of the Ninety-Nines launched a vigorous protest campaign. The CAA backed off but in response imposed a restriction that if during the woman's "recovery" after the birth her license expired, she would have to take the written and flight tests over again. Continued pressure from the Ninety-Nines eventually had this restriction also removed.

"On the other side of the coin," Colleen says, "The everyday atmosphere is friendly and professional. I believe that being able to fly with the others of my generation helps immensely. In the military I haven't had to deal with father-daughter cockpit relationships that I hear about from women friends employed by major airlines. I do not have to constantly prove my abilities and I truly feel that I am respected by my peers and other crewmembers. Fortunately, the Coast Guard has not restricted its women aviators from any mission, unlike the other services. For the most part I have been impressed with the quality of women that make it through aviation training. Like any profession there are always a few who slip through the system, but they are usually weeded out eventually."

US Army

Sally Murphy

About the same time the Navy opened flight training to woman, the Army did so too. Their approach was different. Instead of forming separate classes for its women as the Navy did in the beginning, Lt. Sally Murphy was integrated into Officer Rotary Wing Class 7414 with 24 other men.

On June 4, 1974 Sally Murphy graduated from the U.S. Army Aviation Center at Ft. Rucker, Alabama. She had the distinction of being the first woman to qualify as an aviator in the Army. Lt. Murphy was carrying forward the heritage left by the WASPs.

At the age of 25 and at the beginning of her military career she was of course aware of the distinction but unaware of the far reaching consequences and the sporadic impact on her personal life and career this distinction would hold in the years to come.

Murphy had gone through the same rigorous airplane and helicopter training offered the men applicants. With no previous flight experience she was unprepared for what the training would be like and what the job would hold once the training was completed. Like the first female Naval aviators, the publicity and thrills centered around training and its completion but the real challenges Murphy found of "being first" were faced in the day-to-day work of an officer-pilot.

"Inspite of being the first and only woman, I was allowed to maintain my femininity, but the men did not pamper me or give me special treatment. They maintained a perfect balance in our relationship.[22]

And, like the Navy, there was some hostility and resentment for a variety of reasons. Even today Murphy still finds these reactions although to a lesser degree. "I'm never certain whether attitudes have changed, or whether I've toughened up (probably both). The challenge, however, was to preform to the best of my ability, everyday as an officer and as a pilot. These are the same challenges faced by all lieutenants; by all beginners in their careers."

Murphy says she was fortunate. "My initial assignment combined my military intelligence background with my aviation training. My assignment to the 330th Army Security Agency Company was both professionally and personally challenging, frustrating, and rewarding. A true growth experience. We flew the "real world" of intelligence missions, which was interesting and kept us under pressure to produce results. We flew state-of-the-art aircraft, with oxygen masks, without cabin pressurization on four hour flights. I learned much about the 'glamour' of the job in those first years."

It was at this point that Murphy began to interweave her professional life with her role as a wife, and later, mother. It was a, "A rich, colorful and formative period. One that I will cherish always."

After years of being told that women could not serve in combat arms, many of the women in Army aviation are now integrated into combat arms since Army aviation became a branch in 1984.

Murphy, also a Whirly-Girl, became commander of the 62nd Aviation Company "The Royal Coachmen." The 62nd is a Corps level general support company with thirty-seven helicopters and two airplanes. "The challenges of leading, training, and maintaining this unit of 200 soldiers is an unparalleled challenge and reward," said Murphy. "The unit has an illustrious heritage and a proud tradition. It is a compliment to be chosen to lead these soldiers, and a challenge to try to surpass the unit's good reputation."

Murphy realizes that as she obtains seniority the opportunities to fly will diminish drastically. In April 1991, Lt. Col. Murphy was give a battalion command in Japan. She said, "As I travel in the Army, I am

[22] "Army's First Female Pilot Wins Helicopter Wings." *Army* , July 24, 1974. p.43

reminded that America has provided the best opportunities for women to grow, and we are decades - sometimes light years - more advantaged than other women throughout the world. When I am depressed about a setback, I remember it is our mission to mold the future. I will continue to work to serve as a good example for women in aviation and in other non-traditional careers. The future is bright for all of us."

Marcella Hayes

On November, 1979, Second Lieutenant Marcella A. Hayes earned her Army aviator wings. What made this event noteworthy was the fact that Lt. Hayes was the first and only black woman in the entire U.S. Armed Forces and the 55th woman of the 48,000 officers and warrant officers to graduate from the Army Aviation School.

Lt. Hayes, 23, the second youngest of four girls, grew up in Centralia, Missouri, the "show me state." Although her journey from civilian to Army pilot began in December, 1978 when she enrolled in the Army, in reality it began in college when she enrolled in the Reserve Officers Training School (ROTC) at the University of Wisconsin, where she majored in English. The ROTC instructor talked to her about flight school and it caught her attention.

"I'm kind of adventurous, said Hayes, whose hobbies are mountain climbing, sewing, backgammon, racquetball and fishing. "If there's something there and it's interesting I don't mind trying it at least once."

Well, Lt. Hayes got more adventure than she ever dreamed of. There were five, 1,250 foot qualification parachute jumps to earn her paratroop badge, and a lot of hard work. For nine months she coped with what the Army calls "the toughest training the Army offers." Each morning she attended three to four hours of classes, including weather, aerodynamics, aircraft maintenance, communications, survival courses and navigation. That was just the warm-up for the day. She spent the next six hours on the flight line. "It's rough." said Lt. Hayes. "You keep telling yourself 'I can do it. I can make it through.'"

The toughest part she said was the instrument course. There she had a problem. Her instructor didn't think a woman could handle being a pilot. "I wanted to take her through the instruments. I wanted to see if a woman had any business being in the cockpit," he said. Hayes showed her instructor she had just as much right as the men. She scored an outstanding grade on her final exam and praise from her instructor that was framed in the words, "Intelligent, determined, and a hard worker." Her instructor was no longer a skeptic.

"The drawback in the past said Hayes was that women had a fear of failure. If she can pass the Flight Aptitude Selection Test and the flight physical she can learn to fly, and have the same opportunity as the men." She reminds young women, "You are not expected to know how to fly. The school will teach you that. All you have to do is meet the requirements."

Desert Storm

For the first time in our history American women were deployed to Saudi Arabia in support roles, ostensibly behind the lines as clerks, nurses, military police and hundreds of other military occupational specialties. Women also went to Saudi Arabia as pilots. Their role was to support the men who were to go into combat. This support was to be given from rear echelon positions, or so the public was led to believe. The women pilots were to be used in a wider and more equal role.

On January 16, 1991, The war against Iraq began. On February 24, 1991, the United States Army, Marines and coalition forces launched the largest helicopter assault in history. Part of that offensive were two dozen women from the 101st Airborne Division. They flew some of the more than 300 Chinook, Black Hawk and Huey choppers that lifted more than 2,000 men and supplies inside Iraq in the first hours of the assault. The operation was the first time United States women flew helicopters on an air assault into enemy territory.

"The country now has an appreciation that women in the Gulf campaign perform roles that are just as essential as those performed by men," said Representative John Heinz ,of Pennsylvania.

Army Major Marie Rossi, of Oradell, New Jersey, was one of the first women over the border into Iraq, leading her company of Chinook helicopters of the 18th Aviation Brigade. "I think if you talk to the women ...in the military we see ourselves as soldiers," said Rossi.

"We don't really see it as man versus woman," she said. "What I am doing is no greater or less than the man who is flying next to me or in back of me."

Three times in the early hours of the war, Rossi and her unit, including two other female pilots carried ammunition, jet fuel, and troops into Iraq.

"I lived in a tent with two other female officers. We thought it was pretty neat that we would be across the border before the rest of the battalion," Rossi said.

It was made clear in the newspapers that women, who made up six percent of the Desert Storm troops were advancing with the front. "Women were flying in a combat environment," said Army Major, and Whirly-Girl, Nancy Burt, a former Black Hawk pilot, "not in direct combat."

The law still excludes women from flying Apache or Cobra attack helicopters that seek out and engage the enemy. Female pilots receive the same pay as men and can fly troops and supplies into the battlefield. They also feel the same rush of adrenalin. Says Burt, "You're flying in formation, low and fast, and across terrain that is almost featureless. That gives you a sense of power."

"Flying in lightly armed or unarmed helicopters, the pilots carry only a pistol," said Army Brigadier General Evelyn "Pat" Foote. "If they go down they don't have a lot to protect themselves."

"You try to land the troops as close to their objective as possible," says Lt. Col. Eric Jowers, of the U.S. Aviation Center in Fort Rucker, Alabama, where the helicopter pilots are trained. "Depending on enemy resistance, the choppers could land immediately on the target." At the time of Operation Desert Storm, there were 380 female pilots among the Army's 13,650 aviators.[23]

The ground action in Operation Desert Storm lasted 100 hours, and 129 Americans were killed in action. Major Rossi flew more than a dozen missions during that time, all successfully. On March 1, 1991, President Bush declared victory in the Persian Gulf and ordered a cease fire. Major Rossi was returning in her CH-47 Chinook helicopter from a night reconnaissance in bad weather. Blinded by the weather, she struck a radio tower in Saudi Arabia. She and two crew members were killed in the crash. She was the first woman military pilot to die in Operation Desert Storm.

In mid June, 1991, Congress voted to lift the federal restrictions on women pilots flying in combat and let policy be set by the military organizations. The Bill went to the Senate where it was expected to meet stiff opposition. On July 31, 1991, to the surprise of a well organized opposition, the Senate voted overwhelmingly to lift the restriction on women pilots flying combat. Through their commitment, and performance during the Gulf War, the women pilots had proven to die-hard skeptics that "The Right Stuff" has no gender.

Chapter 7

Astronauts

Firsts are only the means to the end of full equality, not the end itself.

Judith Resnick

In 1983, when Dr. Sally Ride made the historic flight that rocketed her into space, she succeeded in a quest that had started two decades earlier. Her flight had come 22 years, 36 manned missions, and 57 astronauts after the first *Mercury* capsule splashed down in the Atlantic.

The delay between the first *Mercury* mission and Dr. Ride's launch into space was not because of women's lack of interest. *Look* Magazine's February, 1960 issue asked in bold headlines, "Should a girl be first in space?" A photo showed 33 year old automobile test driver, and pilot Betty Skelton dressed in a space suit, posing in front of an *Atlas* Booster rocket. She was visiting Cape Canaveral as a reporter for the magazine, and not as an astronaut candidate. Two thousand American women, mostly teenagers, according to Ms. Skelton, had volunteered for space flight. With such a large number surely some would qualify. In the article, Air Force Brigadier General, Don Flickinger said without elaboration, "Women would not be given serious consideration for space travel until three-person space vehicles were in use."

Flickinger, however was telling a half-truth. He and Dr. W. Randolph Lovelace II, the scientist responsible for medical screening of the *Mercury* astronauts, had already decided to screen some women to see if they were fit for space travel.

In 1959, when NASA selected the first seven male astronauts, they also quietly selected a dozen of the top women pilots in the United States. At the time, the top female pilot was 29-year old Geraldyn "Jerrie" Cobb, the chief pilot for an Oklahoma aircraft company. (Fig. 7-1) Cobb, a fixed wing and helicopter pilot, had more than 7,000 hours flight time, more than any of the male astronauts. She also held a speed and altitude record, and was Aviation's "Woman of the Year" in 1959.[1]

In August, 1960, Lovelace let the secret out at the International Space Symposium in Stockholm, Sweden. *Time* Magazine quickly followed up with a story on Cobb's progress. According to Dr. Lovelace, she had qualified to "live, observe, and do optimal work in the environment of space and return safely to earth." She had added credibility to the theory that many scientists had long believed: women may be better equipped than men for space travel.

"Women have lower body mass, need significantly less oxygen and less food, and may be able to go up in lighter capsules, or exist longer

---

[1] Kathy Keeton, *Women of Tomorrow*. New York, St. Martin's Press/Marek Press.; 1985

Fig. 7-1 Jerri
Cobb

than men on the same supplies. Since women's reproductive organs are internal they should be able to tolerate higher radiation levels," Lovelace reported. One cannot help but be reminded of Professor Rudolph Hesingmuller's 1911 list.

Whirly-Girl Cobb had passed the first battery of tests but she wasn't ready for a space journey. The most grueling tests lay before her: underwater isolation, explosive decompression, a whirling centrifuge, and the oven-hot chamber. *Time* Magazine reported that if she passed all the tests she would, "don a formless spacesuit, tuck her ponytail into a helmet and hop atop a rocket for the long lonely journey into space."

If Americans believed *Time* Magazine, they were in for a disappointment. NASA did not expect a woman to pass the tests so there were no plans to deal with women in the program. NASA was in for a big surprise. Cobb passed all the tests, and NASA now had a problem.

The press approached Cobb's success with the same measured sexism they approached Sally Ride's success twenty three years later, when they asked Ride if she had worn a bra in space. "Bachelor Girl Cobb 36-27-34," read the headlines. A few weeks later Brigadier General Flickinger thought he had a solution to the problem. He said, "The tests were preliminary, Cobb isn't an official astronaut." Then he put his foot in deeper. The women wouldn't go into space because, "NASA didn't have any spacesuits to accommodate their particular biological needs and functions."

The problem persisted. An American woman had qualified for space travel. The officials at NASA had a plan to eliminate the problem permanently. Cobb continued to the next stage of astronaut testing at the

Navy's School of Aviation, in Pensacola, Florida. There Lovelace went through the motions of putting twelve other female applicants through the testing program. He was sure the women would flunk the tests. The women disappointed NASA officials, again, and passed all the tests. Now NASA had a real problem. They had thirteen women fully qualified for space travel and were reeling under public attention to this new discovery. NASA stepped back for a moment and came up with another more permanent solution to the problem. In July, 1961, NASA canceled all testing of women. That should have ended the problem but candidate number thirteen was a Whirly-Girl named Janey Briggs-Hart, wife of then Senator Philip A. Hart, of Michigan. Within a year Hart had a congressional subcommittee investigating the NASA policy. By this time America was deep into the space race with Russia. Yuri Gagarin had become the first human to orbit the earth, and John Glenn was right behind him in early 1962. Glenn went before Hart's subcommittee to explain why NASA wasn't training women. His answer was simple. "Astronauts have to be jet test pilots, and engineers, and the women aren't." When asked to explain the qualifying tests he provided another simple answer. "They're so minimal. As an analogy, my mother probably could pass the pre-season physical given the Washington Redskins, but I don't think she could play many games."

Another NASA official told the committee that training women would, "be a waste, a luxury the nation could not afford if they were to get a man to the moon by 1970." [2]

In June 1963, soon after the hearings, the Russians beat America to the punch when they launched twenty-six year-old Valentine Tereshkova into space. It was political one-upmanship. Tereshkova was by any standard a token woman in space during her 48 orbits in the *Vostok IV.* She was a textile worker and her aviation background was that of an amateur sky diver. Her presence added little more than prestige to the flight. She did, however, go on to become a respected international spokesperson for the Soviet Women's Committee and for Soviet women. Russian chauvinism was the same breed of animal. It would be twenty years before they sent another woman up in space, and then it was to upstage Sally Ride's launch.

In 1964, NASA announced a new policy to recruit an elite corps of scientist-astronauts for the upcoming Apollo and Skylab missions, and they encouraged women to apply. Perhaps this was the breakthrough? Many qualified women applied but not one was chosen, at least not for the next fourteen years.

In 1972, Science Fiction writer Robert Heinlein cleared the smoke. "If you want my personal opinion, women haven't been invited into the space program because the people who set up the rules are prejudiced. I would like to see some qualified women hit NASA under the Civil

---

[2]   ibid.

Rights Act of 1964. They're entitled to go; they're paying half the taxes; it's just as much their program as ours. How in God's name NASA could fail to notice that half the human race is female, I don't know."[3]

Most Americans knew there was much more than a military value to the conquest of space. The commercial and industrial "by products" of space exploration would benefit Americans, and NASA began to focus on a reusable space vehicle to get scientists back and forth to earth.

NASA looked at the shuttle as offering the first real opportunity for women in space. Designed for "everybody" scientists, engineers, technicians, and not just jet test pilots, the shuttle offered room and a minimum amount of privacy. By this time NASA had also taken care of "women's particular biological needs and functions," by building a $3 million zero-gravity toilet to accommodate either sex.

False Start

In 1973, NASA began a new program for testing women to see if they could stand up under the physical and psychological stress of space flight. They apparently misfiled all the data from the earlier tests. Perhaps with this new generation of women there was a mutant gene or an "X" factor. At first Air Force flight nurses volunteered but by 1977, civilian women from all walks of life and ranging in age from 25-65 were test subjects. They were poked and probed, chilled and heated, confined, and cajoled and the results confirmed what the scientists in the late 1950s had found. The women could endure the physical hardships the same as the men, and had better psychological profiles.

The women volunteers did something the men did not do and it surprised the NASA test officials. To endure the boredom and trauma of the tests, the women had formed a tightly knit support group. NASA realized the women had led them to an important discovery. Zero-gravity was a serious equalizer of men and women, and physical strength was not important. We had entered a new age where small groups of people depended on each other as a team. The lone macho jet jock was out. Dr. Harold Sandler, chief of NASA's Ames Biomedical Research Division said, "In space women are going to beat men."

In 1975, current data on female physiological responses to high performance flight was limited. There had been speculation that women may be more predisposed to altitude decompression sickness and may not respond as rapidly to treatment as men. This completely ignored the fact that the WASPs participated in extensive high-altitude flight testing in hyperbaric chambers with no ill effects. It was further hypothesized that for women taking birth control pills this condition could be more serious. Altitude decompression sickness occurs when small bubbles of nitrogen appear in the blood and the platelet count drops as a secondary response. The experience of the WASP program in dealing with the physiology of women in high performance flight testing is particularly

[3] Frank Robinson, "Conversation With Robert Heinlein."

revealing. Medical reasons for elimination of WASPs from the program averaged 35.5%, compared with 35.6% of the male cadets. Principal reasons for medical eliminations in all candidates, male and female were emotional instability and hysteria, air-asthma, claustrophobia and fatigue.

The WASPs were often pushed in training to catch up hours so that classes could be regularly started and graduated. The cases of operational or flying fatigue were so far below the rate of incidence among male pilots that attempts were made to rationalize the results. "Women were so desirous of flying and so determined to show that they were as good as the men they refused to give in to or report fatigue," said one report. Lost time operations for physical reasons for the WASPs never exceeded the same figure for all flying personnel. The WASPs ferry-work lost less time than their male colleagues in the same work.

It was the general belief at the start of the space program that women were handicapped due to menstruation and would be off duty a few days each month, making the regularity of their service consequently undependable. The conclusion of the medical reports of the WASPs that menstruation in properly selected women is not a handicap to flying or dependable performance of duty had also been "forgotten."

It had also been the opinion of many that women lack the muscular strength to do all types of flying. The WASPs flew the Flying Fortresses for more than 12,500 hours with no fatalities and with only three minor accidents. They also flew the even larger B-29. Actually muscular strength is less of a factor in piloting planes, but if it was, that selected women in large numbers are available who have the size and strength for these tasks.

In 1976 some physiological testing had been done in other related areas. Scattered studies indicated that women "appear equal to men for space flight," and increasing evidence pointed to the conclusion that women may be more suitable than men for some missions. The tests on women in the 1960s by NASA had drawn correct conclusions. Women were equal and in some cases better equipped to handle space flight than men.

Women have flown almost every type of aircraft ever built, and there have been women test pilots right along with the men. Jackie Auriol, a noted French test pilot said, "I don't know whether it is because I am a woman or whether there is some other physiological reason, but I have the good fortune always to sense an oxygen failure in time. I am warned by a feeling of nausea. With many other pilots, their minds start wandering before they realize what's happening to them."

In 1976 there were also indications that women may have greater endurance and more stamina than their male colleagues. In modern aircraft physical strength is no longer a major factor. In the event of an emergency where strength would be necessary, it has been argued that the normal physiological adrenaline effect would be the same for women as for men, allowing them to exert several times the strength a female might normally exhibit. Most emergency situations do not require inordinate amounts of strength, anyway. For example, the typical autopilot override mechanism requires only 20-45 pounds of force.

Fig. 7-2 Sally Ride

Sally Kristen Ride
　　Sally Kristen Ride graduated from Stanford University in 1973; she received her Doctorate in physics in 1978, and became an astronaut by answering a help-wanted ad placed by NASA on the Stanford University bulletin board. She signed up and won out over eight thousand other applicants. [4] (Fig. 7-2)
　　In 1978, Dr. Sally Ride was one of six women selected in an American team of 15 pilots, and 20 mission specialists for the upcoming shuttle and space station programs. The other women were Rhea Seddon, Kathryn Sullivan, Anna Fisher, Shannon Luad and Judith Resnick. (Fig. 7-3) In 1979 NASA commissioned a huge mural showing a group of astronauts in the clouds headed by frequent space traveler John Young. Among them was a woman. In the original artist's sketch, made in the mid seventies she was not in the picture. Attitudes were beginning to change.
　　In 1982, then Director Chris Kraft said, "Women have never been excluded from the Astronaut Corps. Whether they were black or yellow, male or female, it made no difference in the selective process. We have

always set the requirements for the people we have needed at a given time, and for a long time women were not trained in those fields." [5]

That's one point of view. But when asked why she thought NASA had decided to recruit women Sally Ride answered, "They felt the push of the women's movement." Ride and others said, "It wasn't a sudden violent push, but a slow, relentless push, like the push of an *Atlas* rocket. It took time to ignite the engine and overcome the inertia of an attitude as ingrained as freedom itself. Once it began to push it would not stop. Its direction was up, toward the stars and victory."

In June, 1983, Dr. Ride flew on the seventh shuttle mission and became the first American woman in space. Sally Ride was an anachronism and a symbol who had no desire in being a symbol. She is an astrophysicist, and a former ranked tennis player. Ironically she never had a deep burning ambition to be an astronaut.

When faced with the fact that a woman was going into space for the first time with four men, reporters asked her if that might create friction. Her reply, "There are right ways and wrong ways to approach anything. We all knew, both the men and the women astronauts, that it was very important for a woman to go up in the shuttle. She would be a role model and that would be important. I didn't particularly care that I was the role model, but I thought it was important that somebody be."

Dr. Ride changed the world above us for millions of Americans. Space was no longer a man's world. Ride did not see it that way. "From the astronaut-office point of view it was not a man's world when I joined NASA along with five other women in 1978. The attitude of both men and women in the program was not whether you were a man or a woman, but could you do the job. The emphasis is on things that work - achievement." When asked why she was picked for the job she answered unabashedly, "Because I knew what I was doing."

During her first space flight someone on the ground asked her to describe the feeling. She answered by first asking "capcom" (capsule communication) had he ever been to Disneyland. Capcom replied, "Affirmative." Ride responded, "Then this is definitely an E ticket," referring to the entrance ticket for the best rides.

Ride also brushed off the celebrity status of being the first American woman in space. "We all knew the first woman would get a lot of attention. It's a tradeoff. There are certain things that go with the job. I get to visit with many interesting people and do interesting things. Given my choice, I'd rather be in a simulator, or better yet, up in space."

By 1984 there were eleven women in the astronaut corps of ninety-two. They were physicians, biochemists, geologists, and ceramic, civil, and environmental engineers. The only requirement now is they be in excellent health with a strong math or science background.

Ride was a nationally ranked tennis player until she decided, "I wasn't that good. I knew it wasn't what I was going to do the rest of my life."

[5]   *GEO,* September, 1982 p.108

Fig. 7-3 America's First Women Astronauts

During her NASA days, Ride gave up tennis for several reasons; for instance, the hassles of scheduling court time, rounding up a friend, and lack of time. Another reason reveals her competitive spirit. "I can't play tennis for fun."

In 1984, Dr. Ride took another ride aboard the shuttle *Challenger,* this time with Dr. Kathryn Sullivan, a geologist. On the eve of her second shuttle mission a reporter asked Dr. Ride a leading question. "Does it mean anything that there will be another woman on board this flight?"

Ride had two answers to the question. "It doesn't matter to the flight. We're both professionals and as a crew we work well together." Ride thought for a moment. "It's also another stride forward for the space program. It proves that women flying into space isn't tokenism, because women are part of the crew." Dr. Sullivan became the first woman to walk in space during that mission. A year later, Dr. Rhea Seddon (M.D.) was part of a team of seven aboard the shuttle *Discovery* sent into space to salvage the disabled communication satellite *Syncom*.

Dr. Ride later became a member of the Presidential Commission that investigated the *Challenger* accident. She went on to become special assistant to the administrator for long range planning at NASA. In the Fall of 1987, Ride left NASA to become a Science Fellow at Stanford University's Center for International Security and Arms Control.

Although at first the press played up America launching women into space, they soon recognized that they were not token women appointed because of their sex. They were the best qualified people for the job.

Before the *Challenger* disaster postponed the shuttle missions, six women accumulated over 1,016 hours in space. [6]

Drs. Judith Resnick and Sally Ride went into space twice. Once the press recovered from its obsessive fascination with the intimate details of the women's grooming, sleeping arrangements, etc., American women in space became unremarkable.

Judith Resnick

At Carnegie Mellon University a plaque in Wean Hall bears the inscription, "My heart is in the Work." These are the words of the Pittsburgh school's founder, Andrew Carnegie. They could be the epitaph of Carnegie's most distinguished engineering graduate, Judith Resnick.

Resnick was born April 5, 1949, and distinguished herself in high school as a classical pianist and the only female member of the mathematics club. She disliked such distinctions.

Judith Resnick earned a Ph.D in electrical engineering and was not a "typical" engineering student. She was a quietly exceptional student intent on maintaining an unexceptional way. She was sharp, articulate, and she did not like attracting attention.

Judith Resnick was also an anachronism. She was a complex person with a reputation for hard work and professionalism. She excelled in competitive professionalism and she didn't match any popular notions of an astronaut. Her private life was no different; still an anachronism. She was the valedictorian at Firestone H.S. in Akron, Ohio, and made perfect scores on her SATs. Although divorced, she continued to share major events in her life with her former husband. She was very close to her father and on her first shuttle flight held up a sign that said, "Hi Dad."

The Challenger Disaster

Two American women in the crew of seven died aboard the space shuttle *Challenger*, when it exploded on launch in 1986. Judith Resnick, a crew member, received little publicity before *Challenger* launched on its last and fatal mission. Equal opportunity in space, and in death, had become a routine operating procedure.

There was, however, exceptional publicity placed on Christa Mc-Auliffe, the first civilian in space. Mrs. McAuliffe was not a token woman on the launch, but a person representing all Americans. The thirty-seven year-old Concord, New Hampshire social studies teacher was chosen from 10,690 educators applying to the "Teacher in Space Project." She was President Reagan's ideal to represent 20th century America. She was going on what she called, "The ultimate field trip."

There was no hint of trouble on the cold January 28th in 1986 when *Challenger* lifted into the aqua-blue Florida sky. Seventy-two seconds into the launch, as *Challenger* was ten miles into the mission and approaching a speed of 2,000 miles an hour, it suddenly exploded. Millions around the world shared America's loss.

The space shuttle marked a historic transition from the days of space exploration to those of space exploitation. One NASA official compared previous *Mercury, Gemini,* and *Apollo* missions to a Lewis and Clark expedition. "We are now sending up settlers. The shuttle is the new Conestoga wagon. Women have always been accepted as pioneers and today's astronaut women represent the steadfast industriousness of the pioneers with the restless curiosity of the explorers."[7]

Mae Jemison

Mae Jemison was only thirteen years old when Apollo 11 landed Neil Armstrong on the moon. That event just reinforced her belief that she

Fig, 7-4 Mae Jemison

could be an astronaut, just like Armstrong. Then too, she had another motivation. Jemison was a avid Star Trek fan. (Fig. 7-4)

"I've wanted to be an astronaut since I was a child," said Mae Jemison. In August, 1987, she began to see her dream come true when she reported to Johnson Space Flight Center. Jemison went on from there to become the first black American woman to qualify as an astronaut.

As a child, Mae, her big sister Ada, and her brother Charles would spend the summer evenings watching stars in the nighttime sky over Morgan Park, a community on Chicago's southern outskirts. They were far enough away from the city lights where the stars were not rivaled by Chicago's lights. In those days Jemison's family could not afford a telescope, so they sat and discussed the seemingly infinite magnitude of space. The fascinations of Jemison's childhood were watching the *Mercury* and *Gemini* launches. She even chronicled the various launches and how they progressed.

Science, space exploration, and astronomy were the kind of things that fascinated young Jemison. Their mother, Dorothy, an English teacher, instilled early in her three children a thirst for learning. Jemison said she never felt any pressure from her parents. "They never stressed good grades, we knew what was expected though." Mrs Jemison encouraged her children to develop their writing skills, and read classical literature. "She felt I leaned too much toward science," Mae said, "so she wanted to round me out."

Her academic regimen helped her do well in Stanford and Cornell Medical College in New York. Dr. Jemison interned at USC Medical Center, in Los Angeles, where she became directly involved with the care of patients. She extended that involvement by becoming an Area Medical Officer in the Peace Corps in Sierra Leone, in West Africa.

It was in Sierra Leone that Dr. Jemison received the application to the astronaut program from NASA. In 1986, Dr. Jemison returned to Los Angeles to work as a general practitioner and take classes at night in engineering. In February, 1987, NASA selected her from among 2,000 applicants.

She describes the interview process at NASA as pretty benign. "I think people have the idea that it's a frightening process. But I think medical school interviews are more difficult," she says, laughing. Dr. Jemison first needed a thorough physical examination, psychiatric interviews, and finally the primary interview. Then she had a background investigation to receive a security clearance. She passed smartly.

After her arrival in Houston, Dr. Jemison completed a year of training before becoming a full-fledged astronaut . The training established a functional "commonality" and a baseline of knowledge, because, though astronauts share "science-based backgrounds," they hail from different disciplines.

Dr. Jemison now works as a liaison between the astronaut office and those who prepare the space shuttle and its cargo for launching. The project involves thousands of people, and there are many tasks that keep Dr. Jemison's schedule a demanding one.

Dr. Jemison is a an honorary member of the Alpha Kappa Alpha Sorority "Because I thought it was an organization that has a very interesting platform of topics." Her other interests include dance, exercise, photography, and drawing.

Dr. Jemison had strong parental support, she said. "They went through all the science fairs and science projects. I had one project where I was doing something with goldfish. My cat ate half the goldfish. It was a regular thing. Once a week the cat would go fishing and my mother and father would go out and buy me more goldfish."

Dr. Jemison points out that there are a number of gender-based assumptions commonly held by parents. "A boy's success in science and math is usually attributed to 'natural ability' while the equal performance of a young girl was viewed as the product of hard work. Not surprisingly, most parents believe education in science for girls is not as important as for boys, for general knowledge or in preparation for careers."

Charlie Jemison, Dr. Jemison's father, believed differently. "My father used to talk to me about how the earth revolved around the sun, about the seasons, about all kinds of things. My father and brother spent a lot of time with me," Dr. Jemison recalls. She notes that when she was growing up, it was considered unusual for a father to exhibit such concern, particularly for a daughter's education.

Dr. Jemison urges children not to "let other people with limited experience and limited ideas about the world" discourage them from pursuing their own dreams. Even the advice of well-intentioned persons, who "may not have the imagination" of the youth they counsel, can be detrimental to a child's aspirations. "My parents never really pushed me any one way or the other," Jemison says. "They wanted me to do well."

Dr. Jemison admits that the astronaut program is something "I don't think just anyone can count on getting into." But she points out that NASA employs geologists, biologists, meteorologists, oceanologist, geographers, engineers, and a range of professionals in science-based fields.

Jemison is training for a trip on the Space Shuttle. Should NASA pursue plans for a moon-based space station as a stepping-stone to other planets, possibly Mars, Jemison expresses no hesitancy about being a crew member. "I would do it," she says. No one in her family nor any of her close friends would be surprised if she went to Mars. The topics of new horizons, and distant galaxies were whispered conversations in her family on many Chicago evenings. [8]

[8] Khalil Abdullah; Grady Wells. "A Galaxy of Expectations." *Sisters.* Spring 1989. p.31

Chapter 8

Commercial Aviation

Being the first was rough. No one was rude to me or did anything, but in the beginning I felt alone. It took about a year and a half before I felt accepted, and it seemed like a long time before I met other women airline pilots.

Capt. Emily Warner

There are many ways a woman can earn a living flying. There is corporate flying, air-charters, air ambulances, showing real estate property, spraying crops, flight instruction, traffic reporting, survey work, pipeline patrol, forestry and air photography. The list goes on but the most glamorous and eventually the highest paying job is flying for the commercial airlines.

In 1987, author Judy Lomax, said, "The United States seems to be the best country for a woman who wants to be a pilot." [1] On a comparative basis with countries in Europe where there are either no anti-discrimination laws or the laws that exist are not taken seriously, the statement is an accurate one. Although the modern gender barrier first fell in Europe, the United States has made giant steps in progress over Europe. Air France dropped the male sex requirement in 1973, but in the next decade, they hired only eleven women as pilots. In 1990 there was only one woman flying for LOT, the Polish airline, and she was hired in 1987. The only Nigerian Airlines female pilot was hired in 1987. The International Society of Women Airline Pilots (ISA) tracks women pilots, and their records back Ms. Lomax' claim. In 1991 there were 1,600 female airline pilots worldwide and over 1,225 are in the United States. (This includes only women who fly under FAR Part 121 with aircraft in excess of 90,000 pounds gross weight). [2]

Although many adults have grown up with the belief that women are equal to men, excluding physical strength, and can do the things a man can do, the image of a pilot for some people comes from some very specific albeit, incorrect ideas of what a pilot looks like. The "right stuff" image appeared on the silver screen decades before Tom Wolfe coined the expression for the title of his book on our first astronauts. The movies usually depicted pilots as tall, square-jawed men, tough and rugged, with military backgrounds of the fighter, bomber, and test pilot mold (like Chuck Yeager). Of course movies like *The Right Stuff* and *Top Gun* went

[1] Judy Lomax. *Women of the Air*. New York. Dodd & Mead, Co.; 1987

[2] International Society of Women Airline Pilots

a long way to reinforce that image. There are exceptions. Actor James Stewart appeared in many flying movies and was also a B-17 bomber pilot during World War II. He went on to rise to the rank of Brigadier General in the Air Force. We know too that many of the real life military pilots usually went on to fly commercial airliners.

We routinely go to women lawyers and physicians and ride with women taxi drivers but do we trust a woman to pilot a commercial airline? Only in the last decade is the answer to that question clear. In 1972, there were no women airline pilots in the United States. By 1982 there were just over 200.

Of the 1,225 women flying for the airlines in the United States today, at least two hundred seventy-five are captains. United Airlines has the most women flying (350). Piedmont Airlines, before it was bought up by USAir in 1989, had the most captains, about 28. Piedmont was also one of the more progressive airlines. On July 10, 1982 Captain Cheryl Peters, First Officer Becky Rose, and Flight Attendants Dolly Wenat and Cindy Perry became the first all-female jet crew on a U.S. scheduled airline. Today USAir continues this "notoriety" with 46 captains of the 150 women who fly for the carrier. American Airlines employs about 140 women pilots, and Continental about 100.[3] But admittedly, with half the population in the United States female, the number of pilots is still small (5 percent of all airline pilots). By the year 2000, one estimate says about fifteen percent of the pilots flying commercial airliners will be women.

For a change, there were no court fights to get the airlines to finally abandon their "men only" practice. When Emily Warner and Bonnie Tiburzi were hired in 1973, the airlines decided then, when they were looking for pilots, that Warner and Tiburzi had the necessary experience and they were qualified. For the women who immediately followed them, their sex was not an issue in the employment office, but in the cockpit.

Although Warner and Tiburzi were qualified, they were still under the male microscope. Airlines maintain rigid training standards and applicants must pass personality and intelligence tests. They also must have an excellent flight record, personal and medical background check, and at least 1,500 hours of flight time.

Getting qualified for the first women entering the field wasn't easy. Usually women entered aviation through civilian training, either through high school or college programs in aeronautics, or private lessons. Until the mid seventies, military aviation training was not open to women and civilian experience was their only avenue. A commercial airline pilot applicant needs experience in advanced aircraft and instruments, multi-engines, and a turbo-jet rating, all available to men in the military and hard to come by and expensive for a civilian. So for Warner, Tiburzi, and many women that followed them it was not simply a matter of acceptance because they had the hours in their log book, it was "let's see if she can really fly."

---

[3] ibid

Doris Langher

For one woman, flying was a life-long passion, and she made it possible for hundreds of men to fly the "friendly skies" of United Airlines. Doris Langher had a low profile career in aviation. She took her first airplane ride in December 1933, and began an amazing forty-year career. Langher went to work for United Airlines as an accounting clerk in Chicago, and soon afterward enrolled in the company's home study courses in navigation, meteorology, and radio direction finding. Flying was in her blood and she soon bought a half interest in a Great Lakes Trainer. She barnstormed part-time, competing in over 100 closed-course races. Her friend Jacqueline Cochran urged her to join the WASPs and Doris almost yielded to the pressure, but she had other goals. She wanted to become the first woman to fly a scheduled airline in "modern times" (She did not count Helen Richey, since her tenure was so short) Langher decided to stay with United Airlines.

When United Airlines moved their Flight Training Center to Denver, Doris followed. By this time she had worked her way out of accounting and into the real world of flying, or at least as close as being a simulator instructor. She also earned a reputation as one of the best multi-engine instrument instructors.

With her busy schedule she found time to fly air races and serve on the FAA Women's Advisory Committee on aviation. To round out her knowledge of aviation she logged more than 12,000 hours with Air Transport Pilot, balloon, glider, and helicopter ratings, (she was also a Whirly-Girl) and this did not include countless thousands of hours in United's jet simulators where she taught hundreds of pilots.

Doris died without ever realizing her goal when her airplane crashed on a training flight while under the control of a trainee. While Doris Langher was not able to break through the cockpit door, Emily Warner made the breakthrough for all women in the jet age.

Emily Warner (1/29/73 Frontier)[4]

Warner's flying career began in 1958, and she was soon instructing at Clinton Aviation in Denver. She soon moved up to the flight school's manager and pilot examiner slots while building her own multi-engine and instrument time. (Fig. 8-1)

The airlines did not go looking for Warner, she went to them, again and again. Warner filed an application with Frontier Airlines in 1967 and updated it regularly. In late 1972, a male flight instructor who worked with Warner said he was going to Frontier Airlines. Warner got mad. She decided enough was enough. She practically camped out on their doorstep for three months. When Frontier opened its doors to hiring, she was there waiting in line, fully qualified. On January 29, 1973, Warner broke the sex barrier in the airlines, or so the newspapers said. Emily

[4] Airline Seniority date

Fig. 8-1 Captain Emily Warner

Warner was the first American woman in modern times to fly for a scheduled airline, and she was proud of that.

Warner reflected on the milestone. "Back in the early 1970s, the airlines' supply of pilots from the military service had started to taper off. So flying jobs were opening up for pilots with other kinds of experience, and my 7,000 hours looked pretty good.

"That's not to say the airlines were eagerly looking for women pilots. I think they were looking for someone to break the ice, to see how the public would respond." The public responded with a deluge of letters. Her passengers applauded the airline and asked, "What took so long?"

The answer to that question is found not in the airline's claim that qualified women had not expressed interest or applied but in economic and image priorities. Before deregulation, airlines could not adjust routes or fares, so they competed for passengers through service framed around image, safety and convenience. The strong, masculine "father image" of the male pilot reinforced this philosophy. At that point too, women had not yet been taken seriously as permanent wage earners. The attitude was still grounded in the 1930s philosophy that Jacqueline Cochran had discussed. "Before she had returned a profit on the heavy investment in such training, she would have converted herself into a wife and mother and stopped working."

Warner started with Frontier as a co-pilot on a de Havilland Twin Otter. After a stint on a Convair 580, she flew her own run as captain on the Twin Otter, and became the first qualified woman captain with a U.S.

airline. By then, she had more than 10,000 hours logged and moved up to co-pilot on a Boeing 737.

As a child Warner was drawn to airplanes and originally planned to be a flight attendant. "When a pilot suggested I take flying lessons, I wondered, can a girl do that? That's how suppressed we were in those days, but I started my lessons the following week."

Warner echoes something most female pilots say. "I've never had any sexual harassment. Flying is so intense and so professional, when you're working you forget the gender of the person sitting next to you. That is not to say there wasn't a sexist attitude, however. My first flight was awful," she said. "The Captain looked at me and said, 'Just don't say anything in the cockpit.' He was the boss."

Warner stayed on when Frontier was purchased by People Express and then by Continental Airlines. She later left Continental for United Parcel Service to fly as a captain on the Boeing 727, and then in May, 1990, she retired. She did not, however, put flying behind her. Today she is an FAA examiner in Colorado.

Bonnie Tiburzi (3/30/73 American Airlines)

Not long after Frontier hired Warner, Bonnie Tiburzi became the second woman to fly for a scheduled airline, American Airlines, and she is considered the first for a major airline (Frontier, at the time they hired

Fig. 8-2 Captain
Bonnie Tiburzi

Warner, had only Regional revenue status.) That was on March 30, 1973, and they are still her employer. Tiburzi was also the first woman to earn a Flight Engineer's rating. (Fig. 8-2)

Bonnie was born into an aviation family. Her father, was a pilot during World War II with the Air Transport Command for TWA. He took her on her first ride when she was twelve. By the age of twenty, she was flying copilot for charter airlines in Europe. Her brother Allen was the youngest pilot to fly the largest transport aircraft at the time, the first stretch DC-8, for Seaboard World Airlines. He became a captain on that aircraft at the age of 31, in 1978. Bonnie's grandfather was also the first man in Sweden to manufacture aircraft parts.

It was not uncommon in the early days to find strong family connections with flying. Women at that time had no visible role models outside the family.

She was also fortunate not to be surrounded by negative influences. "I always wanted to be an airline pilot, and nobody told me I couldn't. One of my first jobs was with a small commuter in Florida, flying a DC-3. We specialized in charter flights for sport fishermen. In those days, a few passengers had doubts about my ability. Some would ask did I know how to get off the runway, or was I waiting for a boy friend.

"I once flew a famous baseball star who demanded to see my license and log book. Then he sat up front the whole trip and kept asking was I sure I could fly. He was a pilot during the war and had never seen a woman fly."

Tiburzi enjoyed the attention that went with her unique status, but soon found herself falling into a trap similar to Helen Richey. "Being the first woman hired by a major airline was great in the beginning. I was single and meeting all these great-looking men. They took me out because I was a novelty. I was on display at social functions and I soon got tired of that."

"Most of the pilots were wonderful and supportive," she said. Some treated her as an interloper. The airline gave her a strength test. She remarked, "Did they think I was going to carry the airplane across the country?" The pilots at American also tested her in other ways. Once they put 10 pound weights in her flight bag, and watched to see if she'd use the "wimp wheels," to pull her bag behind her, she didn't.

There was also some hostility, but not from the pilots. Some people accused her of taking a job away from a man. Others said she was trying to "make a statement." Her reply was simply, "I'm not trying to make a statement, I'm just trying to make a living."

When asked why she flies, Tiburzi's eyes widen, and her mouth breaks into a wide smile. "The joy of this achievement is insurmountable. In the early days," said Tiburzi, women pilots faced issues like whether we should wear skirts or pants, a blouse or shirt or a different hat. Today there are so many more important issues, like whether to fly when pregnant, or how to blend a family and flying, and a career."

Bonnie Tiburzi reminds young women who dream of flying, "If you are determined, confident and committed, you'll do fine. You have to be good."

Tiburzi has received several aviation awards during her career. In 1980, she received the Amelia Earhart award, and in 1974, the Amita Award for the outstanding Italian American. She is also a member of the Wings Club and the International Society of Women Airline Pilots.

JoAnn Osterud (3/10/75 Alaska Airlines)

JoAnn Osterud wanted to fly since she was two years old, but waited until after college to unfold her wings. She had to hold down three jobs and save every penny to take flying lessons and buy her own plane. At twenty-six she worked for a trucking company as a combination pilot/secretary. "None of the other pilots could type," she said.

Osterud became Alaska Airlines first woman pilot and joined United Airlines in 1978. She immediately ran into what every woman in the early days was up against. "At first the guys were really unsure. They were letting women into their private club. But it seems in aviation if you do your job right, eventually people don't really care what you are."

JoAnn is also a professional stunt pilot, and in 1989, she broke a 58 year-old record for flying outside loops set by Dorothy Hester Stenzel, in 1931. Stenzel set the record of 62 outside loops and it stood all these years. One day, Stenzel was watching Osterud fly at an airshow and decided that it was time for a new record holder, and Osterud was the one Stenzel felt could do it. On July 13, 1989, Joann Osterud flew her highly modified *Supernova* biplane through 208 continuous outside loops and set a new official record.

All this is a little ironic today, since JoAnn admits that at one time she "was your original white-knuckle flyer." On her first flight she said she was, "terrified and loving it at the same time." Today JoAnn is a Flight Engineer for United Airlines. She is also a Whirly-Girl scholarship winner,and in her recent past she has flown everything from vintage fighter planes to helicopters.

"You never quit learning, you never learn it all - it never becomes so routine its boring. Besides, I just love to fly. That's why she puts her spare time into Osterud Aviation Airshows.

Angela Masson (09/22/76 American Airlines)

Angela Masson began her commercial flying in 1972 for the Antelope Valley Land Investment Company, in California. She quickly moved up to becoming a flight instructor for the ROTC and on to an instructor's job with Golden West Airways in Santa Monica. (Fig. 8-3)

In 1976 American Airlines hired her as a Flight Engineer on a Boeing 727. By 1979 she had advanced to copilot with American on a variety of aircraft. Today Angela is checked out in every Boeing jet, including the 747, 757, and the 767. She is now a captain with American Airlines but that is not where her interests stop. While flying for American she earned a Ph.D. in Public Administration by submitting her dissertation on,"Elements of Organizational Discrimination: The Air Force Response to Women as Military Pilots."

Like many pilots, Angela flies for relaxation too. Only her idea of relaxation is flying in races, like the Powder Puff Derby, Angel Derby,

Fig. 8-3 Captain Angela Masson

Palms-to-Pines Air Race, Kachina Doll Derby, Hayward Las Vegas Air Race, and the Pacific Air Race to name a few.

Beverley Bass (10/24/76 American Airlines)

It was 5 a.m. December 29, 1986, and American Airlines Flight 417, a Boeing 727, had just been cleared for takeoff. Captain Bass pointed the nose of the jet into the wind, and began the takeoff roll. There was nothing unusual about the takeoff roll or the airplane lifting into the pre-dawn sky, but there was something different about this flight. Captain Beverley Bass (American Airline's first female captain, and at age 35 also their youngest), (Fig. 8-4) First Officer Theresa Clairidge and Flight Engineer Tracy Prior were making history for American Airlines as their first all-female cockpit crew. American Airlines Flight 417 may have the distinction of being both the first and the last; the three women in the cockpit will probably never fly together again. A month later Tracy Prior became First Officer Tracy Prior.

"I wanted to fly ever since I could remember," said Bass. "When I was eight years old, my parents asked me if I had one wish what would it be? I told them I would like to land a jet in San Francisco at midnight." By 1971, Bass was attending college in Fort Worth Texas, and taking flying lessons in her spare time in the afternoons and evenings. After graduation in 1974, Bass had a commercial flying job waiting for her with a Texas mortician. She flew everything in those days, just so she could fly. "I flew corpses for the mortician in an aging single engine Bonanza, loads of freight out of Love Field, airplane parts, and even canceled checks for a film developer. I flew anything and at anytime, usually between 9 p.m. and 3 a.m.. I never saw any women airline pilots and never thought I could ever get the job."

"There were times when I would get pretty discouraged and tired of men telling me 'girls don't fly planes.' Why Not, I asked? The plane

Fig. 8-4 Captain Beverley Bass

doesn't know the difference. I think women have a nicer feel for flying and handle the plane more smoothly. At American Airlines male pilots and passengers have been very receptive. The pilots know I've had to work just as hard if not harder than they did to get where I am today."

There was no discrimination when it came to her test. Bass had to have perfect vision, pass an EEG, an EKG brain scan, and psychological test, and use the flight simulator, just like the men. "Five captains interviewed me. They asked every conceivable question and a few that came out of left field. How much was American Airlines stock? Luckily, I knew. They even asked for the brand name of the propeller on the Bonanza. They were just checking if my flying log was accurate.

"You have to be political too, just like the copilots of forty years ago. If the captain wants to smoke a cigar, and you hate cigars what would you do, one asked me? I'd let him smoke it, I said. I passed with flying colors."

Norah O'Neill (12/01/76 Flying Tigers)

Norah O'Neill is Flying Tiger's first woman pilot. She had four years towards a journalism degree when she switched to a degree in professional aeronautics at Embry-Riddle University, in Florida.

She took her first small plane ride during a ski-clothing modeling assignment in Alaska. The pilot put the film crew and models 10,000 feet up the side of Mt. McKinley, on a glacier, in a ski plane. Five days later, when he returned to pick them up, Norah was overcome with the experience, flying over places untouched by humans, over "bottomless" crevices, slowly grinding glaciers, and places only birds and small plane pilots ever see. It was then she decided to learn to fly. A month later, after taking a cocktail waitress job in Alaska to support her day-time habit, she had her private license. Four months later O'Neill had her Commercial and Instrument ticket and was hooked on flying.

"I discovered that although there was a shortage of pilots for the Alaskan pipeline construction many companies would not even accept an application from a woman, let alone interview one!" O'Neill had several Instructor licenses and started to teach. "Later I discovered that many of my students got jobs in companies that would not even interview me. It was frustrating." O'Neill started teaching at Alaska Central Air, in Fairbanks. They had a flight school, a charter service, and mail runs throughout central Alaska. They also had pipeline contracts to carry men, machinery and supplies, and all the medical evacuations north of the Yukon River. O'Neill hoped that with time and good performance she would move up from the flight school. Not so, again she met discrimination. The airline's twenty-six pilots were embarrassed to work for a company that had a "dangerous girl pilot." The chief pilot even refused to give her a Part 135 check ride. "Eventually I got lucky. A black flight student of mine, no stranger to prejudice, saw what was going on and did something. He turned out to be quite wealthy. He owned several pipeline contracts, and often chartered airplanes. He worked out a long-term charter contract with my company with one clause that I fly his planes. So I became the first woman (and the only one for two years) flying the pipeline. It was my breakthrough. When I left Alaska Central in 1976 to go with Tigers, I was a captain on their largest equipment and regarded as their most reliable pilot. I was not married then as many men were, so I was available for the 2 a.m. medical evacuations, the late Sunday night charters, and long layovers away from home.

"Flying Tigers was my first choice because they were international, with large jets and cargo. I was a little tired of people getting on planes, realizing that their pilot was a girl and getting off, either asking for a refund, or saying they'd wait for the next plane to their destination that had a *real* pilot flying the plane."

Right before her first international passenger trip, several male pilots warned O'Neill to prepare for a bad experience with the old "battle axe flight attendants" who would hate her for entering the cockpit and making three times more money than they did. They were wrong! "For the first time in my aviation career, I was welcomed with open arms! The flight attendants without exception bent over backward to greet me warmly and help. Enroute the senior flight attendant heard the captain yelling at me. She came into the cockpit, pointed her finger at him and said, 'You leave her alone, or I am going to work you over personally when we land.' She slammed the cockpit door on her way out, and he left me alone."

Lynn Rippelmeyer of Continental got the same support from the flight attendants. "As for flight attendants I fly with it's really been touching, and rewarding. They're proud that we have a company that can create an environment that encourages women to be captain."

In 1980 O'Neill checked out as 747 copilot and became the first woman to fly the 747 internationally. (On July 19, 1984, Lynn Rippelmeyer and Beverly Burns were the first women to captain the 747. Burns flew it transcontinently and Rippelmeyer flew it transatlantic. At the time both worked for People Express, and did not want exclusive media coverage of herself. They asked the company to schedule the flights on the same day.

Lennie Sorenson (2/28/77 Continental Airlines)

Lennie Sorenson didn't dream of being a captain on a Boeing 747, and she didn't start out with a burning desire to fly either. After graduating from Ohio University with a degree in Zoology she decided to be a science teacher. As a teacher she could work nine months a year and for the other three, feed the wanderlust that was beginning to nag at her.

There was still more buried inside the Texas-born youngster. The oldest of five children, she felt trapped helping to raise her sisters and brothers. "I could not imagine myself staying in one place for the rest of my life," she said. In the Fall of 1969, at the age of 20, with enough money saved for one semester at the University of Hawaii, she pulled up her roots and bought a one-way ticket to Hawaii.

Once in Hawaii, Sorenson worked hard. She earned a teaching assistant's job which paid her tuition and a small salary. At this point she made a second bold move. She decided to major in Marine Biology, an option that in 1969 was a non-traditional choice for a young woman in Hawaii. As part of her studies she took SCUBA lessons and became a certified diver.

One day on a flight back to Oahu, from a diving trip, Lennie discovered the beauty of flying. The small plane gave her a close-up, birds-eye view of the overwhelming scenery on the islands. She was smittened. When she asked the pilot if flying was as easy as it looked, and as much fun she got an unexpected reply. The pilot looked back at her and said flying was not easy and not many people can do it well. She also sensed something in his voice. "I got the feeling that he didn't like a woman questioning him. I looked him straight in the eyes and thought to myself, he didn't look any smarter than me."

When the plane landed Sorenson walked over to the nearby flight school and signed up for lessons. The flying bug had bitten another victim. Within months she had her Private Pilot's license. She did not realize at the moment but flying was beginning to consume her. She continued her graduate studies for awhile but she was becoming increasingly enraptured by the wings she had strapped on. To support her flying habit, she began to give SCUBA lessons.

During a flight check for an advanced rating, an examiner asked her what she was doing there. So it began, endless questions over the years that she would answer explaining her presence in the cockpit. Her name sometimes caused confusion and sometimes worked to her advantage. It at least allowed her to get her foot in the door. When an interviewer saw "Lennie" on the application he did not think for a minute that the applicant could have been a woman. Her name also acted as a double edged sword. (In grade school they put me in the boy's gym class by mistake." She laughed and thought for a moment. "Too bad I didn't appreciate it at the time."

On job applications she would "forget" to check male or female. "Once I appeared for an interview, and the prospective employer tried to shoo

Fig. 8-5 (L-R) Karlene Chunglo, Captain Lennie Sorenson, & Dorothy Clegg after their historic flight.

me away telling me he was waiting for a man. When I told him I was the applicant he was waiting for, he looked at me and strongly asked me out of his office anyway."

Throughout her career as a professional pilot Sorenson has worked in a fishbowl. "That's just the way it is," she says. "Sometimes the exams were a little stiffer, but that was to be expected. They really needed to know early on that a woman could be part of the team." Sorenson continued, "There are a few people that even if I walked on water would still find something unkind to say. Some people have a hard time dealing with women as pilots." Lennie paused, her eyes were traveling back to some point in the past. "It has taken me a long time to learn how to let them know it's their problem I'm a woman, not mine."

As flying took over her life Sorenson realized that she could get paid for her flying skills. "Once I worked for a company ferrying airplanes. It was one of my favorite jobs. I could log flight time, travel and see the world, and get paid for it too."

One assignment was to ferry a Cessna 182 from California to Australia. She could have said, "No," because she had never flown a Cessna 182 before, it was over a long stretch of water, and it was her and extra fuel tanks - literally a flying bomb. Instead Sorenson said, "I knew I could fly, and if I could start the plane I could fly it. A lot of things could go wrong; engine failure, the fuel might not transfer, I could get lost but I knew I had checked things out well enough, and I also knew the odds were always there."

It was also an incredible challenge. The first leg of the trip, California to Hawaii was over the largest span of water in the world. Navigation was critical. For every one degree of compass heading she might be off in 60 miles, she'd be a mile off her target. She'd have to fly within a degree of compass heading or miss the islands, run out of fuel and disappear forever. In June, 1974, Sorenson landed safely in Brisbane, Australia, and completed her record setting flight. She had become the first woman to solo the Pacific, from California to Australia.

In February 1977, Continental Airlines hired Lennie Sorenson as its first woman pilot. They promoted her to captain in the Boeing 727 in February, 1984. "People get captain's ratings all the time yet everyone knew when I went up for my check ride." It was a big deal to everyone except Sorenson. She looked at it this way. "They let us on the Panel (Flight Engineer's station) and then the window seat (the right seat). But give us responsibility for the airplane? That was a big step forward."

Sorenson reflected. "Back in 1984 I suggested an all-female crew but met some opposition from the men in management. "They said the public was not ready and would not accept it. I told them it was them, not the public. So we had to wait awhile."

She didn't wait long. On January 16, 1986, she captained the first all-female crew on the 727. When her history making flight landed she met the now-quiet opposition with the look a woman may give her husband after taking his new car out of the garage and to the mall. "Hi Honey, I'm back and it is really still in one piece." Sorenson repeated the feat on the DC-10 on August 27, 1987, when she commanded the first all-female crew on the wide-body jet. The three women, Captain Lennie Sorenson, First Officer Dorothy Clegg and Flight Engineer Karlene Chunglo, had no idea their routine flight would gain such media attention. (Fig. 8-5) The Sydney, Australia newspapers all reported the event. It was the first time an all-female flight crew had landed in Australia. The air traffic controller reported the women made a perfect landing, and TV crews filmed every inch of the landing. Sorenson recalled, "It was on all the news stations. Thank goodness I didn't bounce the landing."

Since Sorenson can't SCUBA dive while she's flying she has taken up a new sport to relax and pass the time - Polo, a sport generally considered a "man's game." She has become so skilled at the game that she is now part of Continental Airline's itinerant polo team. A Sultan in Malaysia, impressed with her polo skills, invited her to his country. Did she go? You bet she did. She and her team won, too.

You might say some obvious changes in recent years are women in the cockpit. Other more subtle changes took place and Sorenson was responsible for one at Continental. She and the other women pilots "inherited" the black gabardine uniforms from the male pilots. She suggested the skirt for women but the idea was rejected until 1984, when she became a captain, then, she just started wearing one. The chief pilot noticed and asked, "Is that authorized?" Her response was, "How does it look?" After an exchange of words she said, "Perhaps it ought to be authorized as well as a better hat." Today the ladies at Continental have a choice: slacks, skirt, or culottes -- and high heels. Lennie reminds skeptics, "The airplane doesn't know I'm a female." She thought for another moment, and continued, "And you don't have to look like a man to do "a man's job."

Sorenson handles all tight situations with humor. She answers males who voice resentment to women getting attention, "You know," she says with a warm smile, "if you had given us the vote earlier, this probably would not be a big deal today."

Lennie's persistence and her attitude that she can do anything if she really wants to do it have been the way she has risen above the prejudice. She has also removed the words "I can't" from her vocabulary. Sorenson has advice for young women who are afraid to go after their dreams. "I want kids to know that they don't have to belong to a millionaires family in order to enjoy themselves. If they take "no" and "I can't" out of their vocabulary they have taken the first step."

What does Lennie Sorenson see in her future? More flying, but someday she'd like to break into the pilot sector of airline management. "That's the next ground to be broken for women. I'm sure they will say the pilots aren't ready, just like they said the public wasn't ready for women pilots. But, it's only a matter of time, before one of us gets there."

Holly Mullins (3/21/77 Braniff Airlines)

One of Holly Mullins' first recollections is sitting in the cockpit of an airplane. Holly's father learned to fly during World War II, and her mother and older brother also fly. Her first lessons came at the age of five on trips between Kansas City and Naples, Florida on visits to her paternal grandparents. (Fig. 8-6)

Those positive influences would play an important role in her later life. In her teen years, it became clear to Holly that for some men, a woman in the cockpit was not a good idea. In the ninth grade she was asked to write a career notebook, and interview the chief pilot of a major airline. When he found out she wanted to fly an airliner, he recommended she start out as a stewardess, and learn some foreign languages. This he said would help her "get her foot in the door." He also hinted that she would have to be three times as good as the men and even then there was virtually no chance she'd ever get hired.

Fortunately, she did not accept this advice. When she graduated from college in 1975, she signed up with the Braniff Educational Systems Inc. to study for a Flight Engineer's rating. She finished at the top of her class and had accumulated 500 hours. When she went looking for a job she found out she was over and under qualified. The airlines would not take her because of her low time, and the commuters didn't want her because they knew as soon as she built her time she's be off to an airline job.

Inspite of difficulties, Mullins was able to build her time and in March 1977, Braniff Airlines hired her as a Flight Engineer on their Boeing 727. At the time she was the youngest female airline pilot in the industry (22 years 3 months).

The following year marked a milestone for Holly. She met twenty other women who flew for the airlines and helped form the International Society of Women Airline Pilots. At their first convention she found, "It was wonderful to be around other women that did exactly what I did. I no longer felt like an oddity."

In May, 1982 she and all the other pilots at Braniff were furloughed when the airline went into bankruptcy. She moved on to a commuter

Fig. 8-6 Captain Holly Mullins & dad Don Fulton

called Wings West, in Santa Monica, California, until Federal Express hired her in 1983.

Today, Captain Holly Mullins flies a Boeing 727, for Federal Express and still gets comments. "When are you going to get a real airline job?" some say. But in the beginning the comments weren't as harmless. One flight attendant said bitterly, "You don't deserve to have that job - up there with my husband." Another said, "Women should not be up there with the men." All that is history now for Captain Mullins. She has flown the hours, sharpened her pilot skills, and honed her decision making ability and earned her right to be "up there" in the left seat.

Denise Blankinship (04/01/77 Piedmont Airlines)

Denise Blankinship always wanted to be an airline pilot but was afraid to say so. That was back in the early 1970s. Blankinship majored in interior design and home economics at the University of Georgia. In 1973, the year she got her degree, Emily Warner and Bonnie Tiburzi had just breached the male fortress called the cockpit. Like many other pilots she came from an aviation background but a female airline pilot was just a dream until the Warner breakthrough. Then it was within her grasp.

Blankinship didn't let the opportunity slip from her grasp either. She had her commercial and instrument rating and had about 350 hours. She began to build her time by ferrying planes between Wichita, Kansas, and Scranton, Pennsylvania. (Fig. 8-7)

When the fuel crunch hit in the mid-seventies she and many other pilots found themselves out of work. Her mother and father remained supportive. Her father started flying for Eastern Airlines when she was five and went on to become a captain on the Airbus A-300. He knew the difficult road ahead and advised his daughter, "Fly only if you absolutely love it. Otherwise stay out." With some luck and good timing, Blankinship eventually landed a job in Tifton, Georgia, flying anything and everything she could. "I only had about 550 hours but a lot of determination. I spent the next eighteen months living and breathing airplanes. I worked seven days a week including Christmas, but I loved it."

By 1976 she was flying a Cessna 310 for a small company in Georgia. The hundreds of resumes she had sent out began to get nibbles. The first airline offers came in late 1976, and by then, she had accumulated an impressive log of 3,650 hours, and had passed her ATP and flight engineer's written exam.

The economic and political climate had also changed. The airlines were hiring and the woman's movement was showing progress. Nine airlines contacted her with offers of interviews. Piedmont Airlines was the first and she signed on as a First Officer on March 7, 1977.

Blankinship points out that the public has been generally supportive. "They realized that we have to earn our credentials. We wouldn't be here if we weren't qualified. The airlines couldn't afford it."

The male pilots reaction was also generally the same. "Behind closed cockpit doors," Blankinship points out, "It's the least secure man who hassles a woman. Women who have problems with their crews usually cause the friction themselves," she points out. "It comes from the woman's attitude. If you've come as far as the cockpit you have to accept

Fig. 8-7 Captain Denise Blankinship

that it's still a male-dominated world; learn to live with it."

Cindy Rucker (06/06/77 Western Airlines)

At some point in a person's life a person may look up and see a commercial jet and fantasize about being in the cockpit. Some people keep the fantasy; others decide to make it happen. For Cindy Rucker it was the summer of 1974. She was 28 years old and had a comfortable career as a commercial artist, musician, and part-time flight instructor. Almost at the instant Cindy decided flying a commercial airliner was to be her career objective, she realized time was running out.

"I began immediately without a second thought. Any self-doubt would have weakened the strength of my desire. I decided I would be an airline pilot by the time I was thirty." (She picked thirty as a target because back in 1974 the historical cutoff age for airline hiring was twenty-nine.)

Rucker went blindly toward her goal. Had she known the financial and emotional costs, plus the hardships ahead, she would have probably failed in the attempt, she recalled.

There were many roadblocks ahead that would have tipped off any reasonable person that the goal was unrealistic. The airlines weren't hiring in 1974, and there was a seven-year backlog of highly trained, high-time ex-military jet jocks. But that was only the tip of the iceberg. All she had was a Commercial Pilot and Flight Instructor rating, not even an Instrument rating. She had about a thousand hours in light aircraft, her vision was 20/100, she was ten pounds overweight, and in debt. But things could have been worse, she said, so off to flight school she went.

The following year found Rucker working 16-hour-days flying an average of 120 hours a month. She earned the Instrument, Instrument-Instructor, Multi-engine, and Multi-engine Instructor add-on ratings, Basic and Turbojet Flight Engineer written exams, an Airline Transport Pilot certificate and a DC-3 type rating. She also went deeper in debt.

Rucker said, "You have to go out and work for this job. No one is going to give it to you. The largest obstacle in most people's path is their own self-doubt, fear of failure or fear of taking a risk. The path for the general aviation pilot to an airline seat is probably the most difficult, but it produces the strongest type, individuals who persevere."

The months were slipping by and Cindy was running out of time. She had to develop a method to keep her name and resume active with the airlines. One month she called half the airlines from a list she made up, and sent a follow-up letter to the other half. The following month she'd reverse the procedure. That kept her name always coming up, and showed her enthusiasm. Rucker still had a major problem. She was now twenty-nine and her eyesight was well beyond the limits of the airline requirements. As luck would have it, she happened upon a person who told her about a procedure of vision correction called Orthokeratology. The surgery was expensive. She could not afford it but she really could not afford not to have it, if she wanted to pass the airline eye test.

After the operation Rucker passed the vision test and in June, 1977, Western Airlines hired her just months before the end of her thirtieth year. She had reached her personal goal but her problems were far from over. She realized she was in a unique position when she walked into the classroom on her first day. "I was the only female among a dozen highly qualified professional pilots, most of whom were from the military. I realized that my acceptance as a female pilot in the airline industry depended wholly on my attitude and the manner in which I handled every situation. It terrified me."

Rucker decided there was only one way to survive. She would make the first moves without being offensively aggressive. She would show the men it was okay to feel awkward, angry or confused. She would act with a professionalism equal to theirs, and have a larger sense of humor than theirs. Her method worked, too. By the end of the three week orientation they were working as a team.

Rucker's first trip on the line, a three-day sequence with a layover was almost her last. As the reserve Second Officer on the 737 she had pre-flighted the aircraft and anxiously waited for the arrival of the Captain. "When he arrived he took one look at me, and mumbled something that sounded like an acknowledgement, and ignored me for the remainder of the trip. It was a test of Rucker's endurance and "I felt like I was losing my sense of humor and my endurance. What kept me going was knowing that this trip would not last forever."

The experience was one in a million, and not representative at all. After that flight, Rucker recalled, she flew with the most professional people possible.

"One thing pulled it all together and proved a point." she said "One day we lost an engine shortly after takeoff, an emergency situation feared by the toughest old captains. In the moment of crisis. the traditional limitations society places on us, this role playing as defined by sex, melted away."

During the entire emergency and afterward, her responsibilities and the credibility of her performance were never questioned or challenged. "We became one functioning entity - a crew that all other times are three individual people." [5]

Duana Robinson (2/13/78 Texas International Airlines)
Duana Robinson started flying in high school in the summer of 1973. "I didn't know Emily Warner was hired that year by Frontier. All I knew was that from the time I was very young I wanted to be an airline pilot. The fact that I was a girl didn't bother me." (Fig. 8-8)

Robinson joined PSA (now USAir) in March, 1985, but got her first airline job in February, 1978, at age 21 with Texas International Airlines (later Continental). She flew as First Officer on the DC-9 for five and a half years. In September 1983, Continental filed Chapter 11 and the pilot group went on strike in protest. "I walked the picket line faithfully until January of 1985 when United Airlines called me for training as a flight engineer on the 727. I finished training in January and received my Flight Engineers certificate. United was waiting to finish contract negotiations

Fig. 8-8 Duana Robinson & dad Captain William Bucklin

before putting us on the line so they sent us home. During the time I was picketing, I kept up my skills flying a Citation business jet. In February of 1985, PSA called me." Flying the Citation helped her get hired by PSA. At that time she had 5,400 hours (4,000 hours jet), an Airline Transport Pilot rating in the Citation, and the Flight Engineer's certificate.

Robinson is another example of where a good sense of humor helps. "It is really funny watching the passenger's expressions, especially men when they see me. The first one looks, and can't believe it. He turns and pokes the next guy and points toward the cockpit. That guy in turn pokes the next guy who looks in at us. This was common back in'78 and '79.

"One day I was waiting to use the restroom while we were on the ground. A fellow came out and took one look at me and said that I should be barefoot, pregnant, and in the kitchen, not on an airplane. I replied that my husband was carrying out two of those at home and doing very well. Mostly, you hear "A woman pilot! Boy that's great!" And that's from both men and women. It also helps when I make a good landing. When the people are leaving, and the cockpit door is open a passenger may say "That was a good landing, who made it?" No matter who made it the captain says, 'She did, or 'She makes those kind all the time.' The passenger expected the captain to say, "he made the landing' or 'thanks.' It really throws them.

"Many people not involved in aviation are surprised to find out what I do for a living. It is still a career field uncommon for women and many can't imagine doing anything like it. It is also funny to hear the question over and over again: 'Do you like your job?' I'd like to say just once; No, I hate it, that's why I worked so hard to get here."

Dorothy Vallee (11/29/79 Republic Airlines)

Dorothy Vallee started flying in her junior year at college. She had a $5 coupon for a Discovery Flight and she thought, why not? One flight was all she needed. The flying bug wasted no time on its latest victim. One flight - one bite as the expression goes. At the time, Dorothy did not have a car and used the next available means of transportation to get to flight lessons, a horse.

To earn money for the rest of her lessons she began apprenticing as a mechanic, something others found out she had a natural ability for. Like many people bitten by "the bug" she progressed rapidly through her ratings. In 1980 she won a Whirly-Girl scholarship, which helped her earn her commercial helicopter rating. (Fig. 8-9)

Beside being a Whirly-Girl scholarship winner, Dorothy used to build and fly aerobatic planes. That aspect grew out of her working at the EAA Experimental Aircraft Museum between 1973 and 1975. The first plane she built was an AcroSport, which she flew until 1980 when she sold it. She then built a second plane, a Christen Eagle II which she still owns today. Dorothy has put the aerobatic aspect of flying behind her since the birth of her two children. These days when she's not flying the Christen Eagle straight and level, she flies the right seat of a Northwest Airlines 747-400 and the 757.

Dorothy wants to see more young women enter aviation but cautions her proteges not to be "distracted." "Get involved with flying first. The

Fig. 8-9 Dorothy Vallee

'distractions' like boys don't go away," she said. "They will wait and always be there later."

Pamela Mitchell Stephens (10/03/83 Republic Airlines)

Pamela Stephens has been a Boeing 747 pilot for Northwest Orient Airlines for the past seven years. She flies the international routes from Europe to the Orient. (Fig. 8-10) Before that she was a first officer on a Republic Airlines 727 jet.

Before Republic Airlines, Pam worked as a test pilot for Cessna Aircraft, doing production flight testing of their business *Citation* jet as well as flight testing turboprops. She also did public relations work for Cessna, promoting their learn-to-fly program around the country as their national spokesperson. That position involved speaking engagements and interviews on television, radio, and in newspapers concerning the aviation industry and careers.

Her early flight time came from a small business she started in 1978, about eight months after she got her private license. Deliverance Unltd. was an aircraft ferry service that delivered aircraft all over the world. Her

Fig. 8-10 Pam Stephens

deliveries ranged from a two-seat Cessna 150, from Memphis, Tenn. to Salzburg, Austria, to a Citation jet from Wichita, Kansas, to Japan, the *long way* via Europe, and the Middle East. Since there weren't many women flying the North Atlantic solo at the time she got a few raised eyebrows and curious glances on her fuel stops, especially in Greenland and Iceland. She has made a number of solo trans-ocean ferry flights.

Interestingly enough, Pam began her career in aviation as a flight attendant for United Airlines while she saved the money for lessons and the required flight time to become a commercial pilot. Her first job was with Air Aurora in Sugar Grove, Illinois. It included air ambulance, freight, and passenger charters. In her free time Pam shares flying a Mooney with her husband who is a former Navy Blue Angels pilot. She has a small aviation stationary business and dabbles in aviation art. Her personal goal in life is to see the entire world, before as she puts it, "I kick the bucket."

Amy Correll (4/05/84 Piedmont Airlines)

Amy Correll started flying at the age of 18. She had already completed a year of study in biology, with her eye on a career in veterinary medicine. "I was spending the summer at home and on a perfectly calm evening my father asked me if I'd like to go to the local airport and watch the planes." Watching the students practice takeoffs and landings got Correll interested in trying it herself. (Fig. 8-11)

The next day she took her first lesson. The bug bit her. "I got a job with the flight school, parking and fueling planes, and began working on my private pilot's license. I was jokingly called a 'line boy.' The next year I transferred to a college in New Hampshire with an aviation program. I had decided to become a professional pilot."

Correll earned her flight instructor's rating, and taught flying at the college part-time until graduating in 1981. Still unsure of her future she stayed on as a full-time flight instructor. It was during that time she met

Fig. 8-11 Captain
Amy Correll

her future husband, Jamie Smith. He was starting his aviation career, at the age of 29, as an instructor. Smith left a career in water resource management when the "bug" bit him.

The following year they both got jobs with a local commuter airline as Twin Otter co-pilots. This was Correll's real training ground, flying mountainous terrain in all weather for little money, but she loved it.

"I received extra attention because I was a woman, but never did I feel any discrimination. I was one of two women pilots in the company. If anyone ever remarked about it, I always got the support of my co-workers and boss. My friends and family were very supportive. One day a little old lady came up to ask me, 'Does your mother know what you're doing?' We all had a good laugh later about that."

Correll got a break when a corporation offered her a job as a copilot. "I jumped at the chance. I had a year and a half with the commuter and I wanted to move up. Although the wages were lower than average for the position, I took it with great enthusiasm. I was going to fly a jet! The job was in New Jersey, and Jamie knew I couldn't pass it up. We started the first phase of a long distance relationship."

Corporate flying was not what Correll expected. "The honeymoon with the job was over soon after I completed the flight training. I learned the job involved more catering to the passengers than flying. I also found out it is not unusual for a corporate co-pilot to be responsible for catering the plane and fixing drinks. But not too many of them had to smile when the boss brought his young female companion to the cockpit and said, 'You see, I'm not a chauvinist, I have a female copilot.' Still, it was good experience and it helped my career."

"I was happy when Piedmont offered me a job. The training would be in North Carolina, but it also meant more distance in my relationship with Jamie, who was by then a corporate captain in Newark. I was fortunate because I was able to commute back and forth."

Correll was one of two women in her training class of twenty. "I think the success and good attitudes of the women hired before me has helped the rest of us. I had neither the most nor the least flight experience and I had to work hard and make the grade like everyone else."

Correll found flying with a major airline meant a lot more exposure to the public. "I got daily comments about being a woman pilot. Many people are surprised! That surprises me since there are women astronauts, doctors, and engineers."

Some of Correll's experiences have been humorous. "I was a flight engineer on a 727 out of Boston. During boarding a passenger looked into the cockpit and he was visibly shocked to see me. He gruffly asked the captain what were my qualifications. The captain calmly replied I had a Ph.D. in Astrophysics from MIT. I struggled to keep a straight face while the man took his seat."

Other experiences were not funny. "As a 737 copilot, I flew two trips with women captains. The comments seemed to double. One man expressed surprise that we "two girls" found Chicago by ourselves. Another got off the plane. Occasionally, I did fly with a captain who clearly thought a woman's place was not in the cockpit. Generally, after the first day he could see I knew my job. The work relationship became, if not exactly warm, at least polite and professional.

Correll married Jamie Smith in 1985. He is now a DC-9 copilot with Midway Airlines, based in Chicago. They live near Charlotte, North Carolina, and now he commutes.

Lori Griffith (04/19/84 Piedmont Airlines)

There was always an airplane in Lori Griffith's family. She grew up remembering her mother driving her from flying lessons to Driver's Ed.

She received her Private Pilot rating when she was eighteen and went on to attend Indiana State University on a Talent Grant Scholarship in their aviation program. In two years she completed a four year degree program with a double major in Aviation Administration and Professional Flight. "It was a tough two years filled with summer sessions and heavy course loads. I did nothing but fly, fly, fly." In that time Griffith earned the Commercial, Instrument, Multi-Engine, Seaplane, Rotocraft, Glider, Instructor-Airplane, Instructor Seaplane, Instructor-Instrument, and Instructor-Multi-Engine ratings. (Fig. 8-12)

Fig. 8-12 Captain Lori Griffith

After graduation, she began working for a company flying pipeline patrol of oil and gas lines in Kentucky. Then in August of 1981, Atlantis Airlines, a Commuter in the Southeast hired her.

Because Lori was so young she could not receive her ATP until her 23rd birthday. "I was passed over for my captain upgrade nine times. Finally a captain's position became available in early December, 1983. Both the chief pilot and training officer flew the line for me to hold the slot until my birthday. On that day, I was the youngest female captain in the industry."

Mrs. Lou Sutton, president of Atlantis Airlines said, "Her success is the direct result of her dedication to a goal and the willingness to make the necessary sacrifices to achieve it. Such success should serve as an encouragement to us in achieving our personal goals."

Piedmont Airlines was Lori's first choice and they hired her in March, 1984, as a flight engineer on the Boeing 727. She was on the Panel for a year before upgrading to the right seat. On April 15, 1987 Lori became the youngest passenger-carrying captain in the world when she upgraded to captain on a Fokker F-28 with Piedmont Airlines. She was 26 years old.

Lori has three people to credit for her interest, ability, and advancement in aviation: Her father, her grandfather, and her husband.

"My father and I started flying together. Although my dad had hundreds of hours he started over with me so we could both complete the ratings together. We went to ground school together, took our written tests together, and shared the same airplane for our lessons. Flying was always one of my father's many interests; he did it with me hoping I would share his interest and ambition.

"The airplane, a Cessna 172, was my grandfathers and he loved to fly. My first airplane ride was with him when I was only six months old. I think of him often and I'm sorry he isn't around now, he'd be proud of my accomplishments." She paused, smiled, and continued. "He would have been the proudest.

"My flight instructor asked me one day why I was attending Indiana State University since I was from Illinois? When I told him about the scholarship he put two and two together and found that he was on the board that decided the recipient. The selection was down to two and he cast the deciding vote. We kid that he chose his destiny when he selected me. Little did he know we would meet, fly together, and marry."

Lori's husband Gregg has been an air traffic controller for the past eight years. "We enjoy talking to one another in the air and share interests in the industry we both love."

Although her husband's degrees are in the same field, when they started making plans to marry they knew it would be a difficult life if both of them tried making a career of flying. "He always had an interest in air traffic control so he chose to be the one who stayed on the ground. His flying background proved an asset on the day he talked down a lost and disorientated private pilot from a thunderstorm. He worked with him for over an hour, and stayed with him until he landed safety. I've never been more proud of him than on that day. He has encouraged me and supported me throughout my career and is a true professional."

Lori has had several all-female flight crews and she finds they are still a novelty to many passengers. "I've had countless hilarious experiences and I've learned that it helps to have a good sense of humor. The best remarks come from the innocent people who have no inclination that you could possibly be a pilot. Once while based in Myrtle Beach, S.C. I helped a little old lady onto the airplane. Before making my way to the copilot's seat, I stopped to help her with her bags and made sure she had her seat belt on. When we landed she remarked as I helped her out of the airplane that it was nice that the captain let me sit right up front! She obviously thought I was a lucky Flight Attendant with a front row seat.

"There have also been people who have asked me for something to drink or to hang up their coat, not realizing I was their captain. Once a woman ran to the galley as we were taxiing out to the runway to ask the flight attendant why we were taxiing when it was obvious that the pilots were left at the gate. We had an all female crew and I'm not sure just what she thought we were doing up there.

"Whether you're male or female, it makes little difference," said Lori. "It takes a certain type of personality for this job. Adaptability is a main character strength. You have to be able to deal with a schedule that changes monthly. Forget about those nine-to-five jobs. You also have to understand and accept that you'll miss some things like Christmas and other special occasions.

Denna Gollner (6/24/85 United Airlines)

Denna took an introductory flying course offered at the University of Colorado. It was something she had always wanted to do and like many others, it hooked her. "I'd always wanted to fly," she says. "I went through ground school, then 12 hours of basic training. I was addicted - I couldn't quit." She then transferred from the University of Colorado where she was a Biology major, to Embry-Riddle Aeronautical University graduating with a B.S. in Aeronautical Science. There were few women enrolled at the university when Gollner was there. "The ratio of men to women was about 50 to 1," she recalls, "but we were all buddies."

"The biggest problem," she said, "was when I got out of college in 1981. Many pilots who were furloughed from the airlines had filtered down into other aviation jobs, so I looked into flying for the military.

"I went into the USAF Reserves shortly after graduation from Embry-Riddle, building up hours flying for the Reserves." She also worked part-time as a flight instructor.

Gollner has flown everything from small single and multi-engine prop aircraft to the T-37, T-38 jets, the C-141, and the DC-10. She is currently a first officer for United Airlines.

"Most of my colleagues treat me professionally," Gollner said. "When they make sexist remarks, I just take it in stride, even if they are serious. It's amusing that no one ever thinks I'm a pilot when I tell them that I fly - they automatically think that I'm a flight attendant or navigator in the military. I've seen a lot of shocked faces when they find out otherwise. Several times when I've been in the cabin when passengers are boarding, they come to me with seat problems, etc., because I'm female.

Gloner's advice to young women - "Always be professional. You still have to be better than most guys to be considered an equal although that is slowly changing. I've seen a few females damage the reputation of many when they take advantage of the fact that they are female in a crew member environment. Gender should not be a factor, but sometimes it still is."

Gollner, like many women, put her career ahead of her personal life and made some sacrifices. The nature of her job makes it hard to make long term plans and her schedule doesn't mesh very well with the 9-5, weekends and holidays-off life. "However," she said, "that's one of the things I enjoy most about flying - it's varied."

Ann Singer (06/29/84 American Airlines)

In 1977, Ann Singer took her first ride in a small airplane with a friend. The feeling of freedom and being able to fly just hundreds of feet above the trees on a spectacular fall day in New England enchanted her. Later, when she had the hands-on sensation of flight, she was hooked.

Ann's college did not offer aviation courses so she squeezed in the time for flying around her course schedule. A shortage of funds was her first challenge. She stopped training several times while she saved to pay for additional lessons.

In 1978, Jim Rowley, Ann's instructor, introduced her to the possibility of becoming an airline pilot. He showed her an article that mentioned

Captain Ann Singer

approximately fifty women flying for the major airlines. "The total percentage of women airline pilots at the time was extremely small," said Singer, "but Jim provided encouragement and remained steadfast in his conviction that if I worked hard I could get hired by an airline. He became my mentor, and good friend. He helped me believe I could do it, and was my source of inspiration. I could never do enough to honor his memory."

After graduating college Ann got her first aviation job, dispatching aircraft and scheduling the pilots (as well as handling all the bookkeeping, clerical and telephone lines) for a seaplane charter and FBO.

Ann did flight instructing and flew company personnel on corporate flights to build flight time. "I was fortunate to have a supportive chief pilot, Tony Acosta, who allowed me to progress up the ranks as a pilot for the company," she said.

Soon Ann was flying charters in amphibious and float-equipped aircraft from Maine to Virginia. Some of her most memorable experiences were flying under the bridges (when it was legal) into Manhattan's East River and docking at Wall St.

After several years flying float planes Ann moved up to a job with Newair, a commuter airline. This gave her the instrument, multi-engine, and high-density airport experience preferred by the airlines.

After six months with Newair, and at the age of 23, Ann earned her captain's stripes. A year later, American Airlines hired Singer. Her determination had made her dream come true.

After a year as a flight engineer on a B-727, Ann upgraded to copilot on a DC-10, flying the international routes. In those days, the captain would often remark that it was the first time he had ever flown with a woman. "For many of the senior pilots having me as a 25-year-old copilot was probably more of an oddity than my being female," she said.

On New Years Eve 1989, Ann completed her captain's training on the MD-80 series aircraft. Captain Ann Singer has found the response from crews and passengers has been overwhelmingly positive. "The advantage for women in choosing a career as an airline pilot," said Ann, "is that advancement is based on date of hire and on passing the required tests, and not on company politics or gender."

Janis Keown Blackburn (3/28/85 Eastern Airlines)

Janis Blackburn worked as a Second Officer on an Eastern Airlines Airbus A-300. According to Airbus Corp., she is the first woman in the world to be crew member on that airplane. "I was thirty-six years old when Eastern hired me. To get hired and to be the "first" at something was exciting," she said. "I'm older than most of the other girls, I'm about the same age or a few years younger than the First Officers. I have more in common with them." (Most of the second officers are in their twenties.) "We talk about the same things in the cockpit. My daughter is in college and many of the First Officers have kids in college. So we have the same problems, tuition, etc. If you're single and 23 versus married with a family you have different things in common." (Fig. 8-13)

At age 14 Janis had her first ride as a Civil Air Patrol Cadet and immediately fell in love with flying. "I knew then I wanted to fly." On March 9, 1968, Janis soloed two days after her 20th birthday, and on May 15, 1969, Janis got her private license. In 1977, she earned her Instructor's rating, and then taught for 3 1/2 years.

A year earlier, Janis flew in the last Powder Puff Derby. "I remember hearing on the news when I was very small (maybe five or six) that, 'Today the Powder Puffers began their air race...' It was then that I decided I wanted to be in one. I didn't know then you had to be a pilot to enter.

"It was exciting to be a contestant in that race. A friend worked for a company that agreed to sponsor us. The company went bankrupt two weeks before we left for California, so my co-pilot, Claire Korica, and I paid our own way.

"When we began to taxi out we saw hundreds of spectators and in the from row I saw my six year old daughter Sandra. I recognized her in the crowd because she was wearing a bicentennial dress her grandmother had made for her.

Fig. 8-13 Janis Blackburn

"We didn't finish in the top ten but the object was to finish the race and we did that. We were flying my Mooney and I had no intention of damaging anything to try to win an air race. I learned a lot about flying and have a great scrapbook and photo album full of memories."

Janis flew for Princeton Airways for 15 months until they went out of business. Then for a year and a half she flew charters until Summit Airlines hired her to fly freight from Philadelphia. "When I flew freight I was away a lot. When I was home it was not quality time with my family because flying freight is a night job, and my "clock" was totally messed up." She stayed for 18 months until Sun Country Airlines offered her a job in Minneapolis, as a second officer on a Boeing 727. When she arrived at their headquarters they decided to train her as a standby First Officer. She flew both right seat and back seat for five months, until Eastern called. "In five months I got home approximately 20 days. I missed a lot of my daughter's early teens."

Janis always wanted to fly with Eastern. She grew up just outside of Philadelphia and TV advertising always featured Eastern. "I applied to Eastern for eight years before they called. The personnel officer said, 'I've been reading your name for eight years and I'm getting tired of it.'" Janis replied, "Then your going to have to hire me because I'm going to keep on writing."

When Janis first applied to EAL she had about 1,500 hours and they were looking for 2,500-3,000 hours. Viet Nam was beginning to shut down and the Air Force was discharging pilots. The airlines could hold out for the pilots with the jet experience, and Janis really didn't have any back then.

Like most pilots Janis had personal struggles. Her Flight Engineer's ticket cost $7,000 and 7 weeks away from her family, and flying with low seniority takes a toll on the family. Airplanes fly 365 days a year. "One Christmas away," Janis relates, "a senior flight attendant bought little Christmas presents for everybody in the crew. We were like family. None of us were with our families, there were nine of us together just enjoying each other's company."

Janis is no stranger to public reaction to a female pilot. "When I walk to the rear I'll get, 'Excuse me honey.' It's always honey - can you help me with this? Sometimes I'll help. The men especially, 25-27 years old are funny. They'll look at me and say, 'you're a girl!' And I'll say, I think you're right. Between the reaction from some of the guys you fly with, some of the passengers and some of the flight attendants, you have to have a good sense of humor."

A male flight attendant went up to Janis one day and said it was wonderful because now the flight attendants could get the pilots pregnant. "I had a new flight attendant come up and call me ma'am - that didn't go over well with me. Another came up and sat in my lap. He thought I was cute."

Janis is still a member of Civil Air Patrol and tries to return what was given to her by the CAP long ago. Each year the New Jersey Wing holds a yearly solo school for cadets. Janis has taught at nine of them. "One of my biggest thrills was watching my 16 year-old daughter solo, in 1986. I recalled every minute of mine. In high school I was always saying that I wanted my pilot's license. Everyone thought I was crazy. For our yearbook friends would write something about you under your photo. Mine says, 'I want my pilot's license.' When I went to my 20th reunion, I won the prize for the most unusual career. Everyone was shocked and most people thought what I did was "pretty neat."

"Quite often, while walking through an airport terminal, I'd see people stop, turn, and watch me walk past. It must be because I'm a pilot; I'm no Loni Anderson! Males in their 20s will sometimes do a double-take when I step out of the cockpit. Once we had a mechanical delay. The captain got on the PA in his 'captain's' voice and gave an explanation that only another pilot would understand. A few minutes later I stepped into the galley to get coffee. A lady stopped me and asked about the problem. I explained in simple terms, referring as best as I could back to an automobile. Then I mentioned there were three very important people on the airplane, and we certainly wouldn't endanger them. She looked around and asked who? I told her the captain, the first officer, and ME! She laughed, nodded, got a magazine, and took her seat."

There's another part of flying that Janis loves. "I enjoy the walk-around on the aircraft. No matter who bought it, while I'm there alone with it, it's *mine!* I pat it and talk to it." Janis flew all of Eastern Airline's A-300s, and each had it's own personality. "They were like children, she said. "A naughty one will give you little problems every time you fly it (nothing serious, just irritating). I love the few minutes I have alone to look at the beauty of the plane."

Janis admits that it's not all wonderful. There are long hours, many late nights, weekends, working on holidays, and jet-lag. But, she says, "as the aircraft lifts off, and all the beauty and spectacular feelings take over everything else is forgotten. I can't imagine doing anything else."

On February 27, 1989 Janis stepped off an EAL A-300 for the last time. On March 4, 1989, Janis and 98 percent of her fellow Eastern pilots chose to honor the machinist's picket line. During the 264 day strike, Janis worked the Newark strike communications office five to seven days a week. In July of 1989, she was elected Second Officer representative for the New York base to the Master Executive Council (MEC) of ALPA. She was the first female on Eastern Airline's MEC. To help pay the bills during the strike, Janis took odd jobs, cut trees, was a movie extra and helped with the census.

"If you do your job, are knowledgeable, pay attention, and do what you're paid to do, you're accepted. This goes for both female and male pilots. The guys I work with resent pilots (either sex) who act as though they've never seen a plane before and don't know what to do with the one they're sitting in. Captains do not like surprises, so informing them of what's going on keeps everybody happy. Back in the early days if I flew with a guy who felt I didn't belong there, I worked a little harder. Now I just work hard because it's ingrained. Besides, that's what I get paid for."

Kathleen Sullivan (08/08/85 Piedmont Airlines)

Kathleen Sullivan is one of five children from Sioux Falls, S.D. One day she went flying with a friend and told her parents she'd like to take lessons. They said she could do whatever she wanted with her own money. Since she didn't have any money she dropped the idea. She went to college for a year but quit to start showing horses to save her money. (Fig. 8-14)

About three years later, Kathy went back to school. Her parents said she could go to any school she wanted and they'd pay her way. Sullivan went to Liberty Baptist College, in VA., and majored in Business Administration with a minor in Aeronautical Science.

"It's funny how I got started. I ended up in the wrong class, a ground school, and I was too embarrassed to leave. The teacher saw me, the only female in his class, and talked me into staying in the course. He took me out for an introductory flight. Twice we went out and each time the airplane was broken. The third time it was still broken and I was beginning to lose interest and wondered if this was meant to be. He noticed that and borrowed a Piper Aztec. He did turns and maneuvers and let me take the controls. He said I was a natural and convinced me to stay with it." Kathy soloed in seven hours and picked up a private, commercial, instrument, and instructor ratings the same year. Then she started teaching for the school, got a part-time job as flight controller, and learned about the commuter business. She earned her multi-engine rating and did a lot of pick-up work for sale airplanes. "The first time they sent me to Baltimore I didn't even know where it was."

"I was teaching 10 students a semester, getting a lot of flight time but my social life was nil. I was up at 5 a.m. and flying while it was still dark to get a few hours in before my 8:00 a.m. classes. It was a tight schedule even in good weather. I stopped around nine or ten at night to go home and take care of my own studies. Weekends were more of the same."

Sullivan had a job offer from Tennessee Airways before school ended, but she turned it down because they wanted her before exams. She graduated college in 1983 with 1,000 hours and sent out her resume.

Atlantis Airlines hired Sullivan to fly the 19-passenger Jetstream turboprop. She stayed with them for eight months and then went to Eastern Metro Express which at the time had six pilots. She talked them into hiring her with only 1,300 hours. "I flew the right seat for three months and then upgraded to captain on the airplane. Metro went from two airplanes to 14 and hired dozens of pilots. I was number seven."

"The move over to the left seat scared me to death. I never had that type of responsibility, and it terrified me. I had always deferred to the captain. Suddenly I was the boss; a role reversal. I could fly the airplane like putting on a glove but as captain the first two or three weeks were very hard. I wanted to go back to the right seat but some of the male pilots told me to hang in for another month. I did and became very comfortable with command. It was the newness that had scared me."

Sullivan stayed with Eastern Metro Express 14 months before lining up a job with Ozark Airlines. She tells a story of the job interviewer's criticism about her school and her convictions. "I graduated from Jerry Falwell's school and had to listen to the interviewer's harassment. I finally

Fig. 8-14
Kathy
Sullivan

stood up and said I was not going to be embarrassed about my school. If he didn't like it that was fine but those were my beliefs. He surprised me by shaking my hand and saying he belonged to the moral majority and was testing my beliefs. He wanted to see if I would wimp out. I didn't, and he hired me."

After she gave her notice to Metro, Sullivan got a call from a friend at Piedmont Airlines. "The two of us went to lunch with the chief pilot, and since I didn't approach him about a job he took my resume back with him. I interviewed with them and they hired me. As long as you do the job, and carry your weight people won't harass you. Come in with a chip on your shoulder and you're in big trouble. I've just never had anybody give me a bad time. Come in with an attitude like hey, I know it all, don't tell me, then you'll have a problem. I figure some day I'll be in that left seat and if there's something I don't know I'm never going to be too proud to learn. There are things you can't learn in the book and a lot of guys know a lot of those things. If you're willing to learn, if your curious they're going to show you."

In December 1986, Sullivan moved up to the Boeing 737-300. "I like to do a lot of hand flying from take-off to 18,000 feet but with computer assist. I like to fly and the airplane has so much capability. It has a little flavor also, it's not boring. That's what I like about flying, its always changing. I don't mind when weather gets bad, it's more of a challenge and you have to be more careful, your battling elements not only flying an airplane." Sullivan has flown with all-female flight crews and she is surprised sometimes by passenger reaction, "Once the passengers gave us a standing ovation. It's funny how some people react. I would walk through the terminal with the guys, and people would look at me. But when there are two or three gals it's unbelievable the looks you get. Once a woman said, 'you made me proud to be a woman.' She made me feel good when she shared that with me.

"Public acceptance of us has been generally good. I had a man ask me three times was I really up front. Finally he said to the flight attendant "Oh, Lord I need a double martini." I haven't seen anybody get off a plane because of a female pilot. At least none that I know of."

Sullivan echoes something most women know. "When you're a female you try harder. I'm competitive. I'm not trying to out-fly the guys, I just want to do the best job I can, make the best landing, and fly as smoothly as I can. You have to know it as well as they do. I have a good rapport with the guys and gals. I really love my job." In her spare time Kathy raises horses and does a lot of riding.

Norma Jean Thompson (04/08/86 Frontier Airlines)

Jeannie Thompson began her career in aviation at the age of seventeen, behind a desk at an airport, writing fuel tickets and answering phones. (Fig.8-15) She actually started flying at the age of 23 when she took an introductory flight at the Cessna Pilot Center. "From that moment I knew that I had to at least solo. I did not think I'd ever be able to afford the commercial ratings but I also believed that where there's a will, there's a way. I ate, slept, and breathed flying. My love and need to be part of the world of flying kept me going."

Fig. 8-15 Jeannie
Thompson

Jeannie was close to being right about not being able to afford the
ratings. She had to stop flying several times because of finances. And
then she ran into the age-old bug-a-boo. She had been furloughed from
a small company when they changed aircraft type. When they decided to
call her back they found out she had a husband (to support her they said),
and the company brought back a man below her on the seniority list first,
since his wife was not working.

For Jeannie Thompson, the flying bug chewed away at her slowly, and
by 1980, she was flying for Air Virginia, a commuter airline. She didn't
realize at the time, but she was hooked. She was driving two and a half
hours to the airport every time she had a flight.

After earning her commercial and instrument ratings she taught for
several years, "still really having fun," she says, "with no intentions of
making flying a career." Around 1978, Jeannie had an experience typical
for the time. In her first job interview with a chief pilot he asked her why
at 28 years old she had so little flight time. She answered that she had
started at 23, and stopped to stay home with her newborn daughter. He
responded that he had five children and that was no excuse for him. She
didn't get the job that time around. A year later she tried again and this
time the company said they were hiring pilots from a defunct airline. "The
frustrating thing was that I was never given the opportunity even though
I was more qualified than some of the pilots they were hiring. Thompson
chalked that experience up to a "good ol' boy" attitude of one chief pilot.

She does admit too that, "Once my colleagues knew I was knowledgeable and capable I was well-received." Occasionally she ran into the
attitude, "She'd make a cute little copilot. I just hope she doesn't want to
be a captain." Her opposition didn't just come from her fellow pilots.

"One day I had the wife of this crusty old captain come up to me and say, "I sure hope you're in the right seat, and stay there!"

But when Jeannie started flying times were different. "I remember feeling on display (and sometimes still do). She has countless stories about the early days when women were a rare sight in the cockpit. I was flying captain with a female copilot and we had five Arab businessmen get on in Philadelphia. They looked into the cockpit and together got off the plane. They refused to fly with two females!"

In the early Eighties, Thompson moved up to corporate flying on a Falcon Jet. The company she worked for applied their own subtle pressure. In their twenty-five-year history they had never hired a female pilot and they made a point of reminding her of this from time to time. It was actually her "single status" that bothered them. One day, they called her into the office and questioned her extensively. "They wanted to be sure I was *settled*, in their words."

In 1986 Frontier Airlines hired Thompson. She was very happy to have what she considered her first real airline job on an airline that in their forty year history never scratched a passenger, even with the challenge of the Rocky Mountains. Her excitement was short-lived, however, when Frontier went bankrupt a few months later.

She enjoyed the job and when Continental Airlines bought up Frontier they offered her a pilot's job on a Boeing 737. After 16 years of flying Jeannie Thompson feels she is where she belongs, something she never dreamed of when she first soloed in 1972.

"The bottom line is I was going to be here one way or the other. Times are changing but a woman still has to believe in herself. Self confidence is the most important characteristic for a woman. There are still not a lot of mentors for girls to admire, therefore one must believe in herself even when the going gets rough."

Someone once asked her when she started taking her career goals seriously. Her answer; "I always strive to be the best, most professional pilot but beyond that, flying is my first love, then my career. I fly simply because, I love it."

Lucy B. Young (06/05/86 Piedmont Airlines)
Lucy Young won a four-year Navy ROTC scholarship to Purdue University and graduated with a B.S. degree, in 1976. After graduation she received her commission as an Ensign and reported to Naval Aviation Schools Command, in Pensacola, Florida, in October 1976. A year later she had earned her Naval aviator wings and then went to jet transition at the Naval Air Station (NAS) Kingsville, Texas.

After qualifying in the TA-4J Skyhawk, Young reported to NAS Barbers Point, Hawaii, where she accumulated over 1,000 hours in fleet support missions and multi-national exercises. Young qualified as a Section Leader, Instructor Pilot and Air Combat Maneuvering Pilot, and then received orders to Training Squadron, NAS Kingsville, Texas. As a Flight Instructor, she taught student naval aviators in all phases of advanced strike training. Young was also one of the first women in the Navy to carrier-qualify, in May, 1982.

After leaving active duty in July 1983, Young accepted a commission in the Naval Reserve. In December, 1983, she accepted a position in Atlanta, Georgia, as the first female FAA Flight Test Pilot. There she performed flight tests on aircraft, avionics and navigation equipment for FAA certification.

In June 1986, she began training to become a Boeing 727 Flight Engineer, with Piedmont Airlines, and is now flying the line with USAir. As a Reserve Naval Aviator at NAS Atlanta, Georgia, she is an Aircraft Commander in the C-9 aircraft.

Lucy is a member of the Ninety-Nines, Tailhook Association, and Society of Women in Military Aviation. She has over 4,000 hours logged in 40 different aircraft.

Camela Condon (07/02/86 Piedmont Airlines)

Camela Condon got started in aviation in 1978. "I always wanted to fly. I can remember laying in the backyard in my grandparent's hammock watching planes go over and thinking how wonderful it would be if I could be the person taking the planes to their destinations. Well, I put the dream in the closet as I grew older."

In college Condon took a Private Pilot course. "I was very excited that my childhood fantasy of flying might become a reality." Condon was so excited about the course she made her mother drive to school in the middle of a terrible snow storm so that she would not miss the first day and possibly her spot in the class. She spent four months in ground school before she ever got in the air. But once in the air she had no doubt what she wanted to do for a living.

Condon graduated college in 1981, with a B.S. in Communications and an Associate in Aviation, and 600 flight hours, her multi-engine-commercial, and instrument ratings, but no job prospects. "The only thing I was sure of was that I would teach flying only as a last resort. I was never comfortable as a instructor. I felt there was so much for me to learn; how could I ever try to teach anyone else."

Condon graduated at just about the time deregulation was expanding the airline business. She landed a job with a commuter airline in her hometown just three weeks out of college. Condon started in a Piper Navajo, then flew an Embraer Bandeirante (Bandit) for six months before moving up to a Shorts 330. Condon then upgraded to captain on the "Bandit" and flew that for six months before deciding to move on. Condon wanted to fly for a major carrier and needed jet time.

Moving up into jets was not going to be easy, Condon realized. She applied to over a hundred corporations and got less than a dozen responses. One of those responses landed her a job flying a Falcon 202 for a corporation in Dayton, Ohio. The job was interesting, she learned a lot but she still had not gotten to her final goal, a major carrier. It took a year and a half of mailing resumes two to three times a month to every airline before she received a response.

"Piedmont Airlines was my first choice, and the first one to call. It was eight years from the day I took my first airplane ride to when Piedmont invited me to come to class, but every step to get there was worth it."

Condon admits there have been many minor instances where she has run into problems because she is a woman pilot. "People love to say the dumbest things and think they are cute, at my expense. Some of the situations I have handled, some my flying companions have helped me with, and some perfect strangers came to my rescue. Some were left hanging because there was little to do or say that would have been professional.

"One day while working for a commuter airline I was doing some paper work at the gate. A man walked up to the counter and said to the agent 'If she's going, I ain't!' The ticket agent turned the man over to me. I told him I was well trained and that my company would like to take him to his destination, but without him we had six other people going. Without me nobody went. I went, he did not.

"Coming into one airport I made a bumpy landing. It happens to everyone. The cockpit door was open and as a passenger got off he stuck his head in and said 'I see they let you make that one, honey.' I felt two inches tall. Before I had a chance to recover the captain said, 'That was one of my better landings, why? Didn't you like it?' The man immediately shut up and walked off the airplane.

"A perfect stranger stuck up for me in Baltimore. As I walked through the gate a passenger in the waiting area started saying how glad he was that I was getting off his flight, and that he didn't want a woman in the cockpit of his plane. I ignored him and continued to walk down the terminal. The man yelled stupid comments after me. Finally he yelled, 'The next thing you know we're going to have lady plumbers!' The man next to him finally stood up and said in an equally loud voice, 'I'd be happy to have her fly me, and if you didn't upset her nothing can.' The first man shut up."

Some comments don't deserve a response, Condon points out. "If God had meant for women to fly the sky would be pink. Can you fly every week of the month?' Or the classic, 'Quit this job, get married, have babies, then you'll have a happy life.' At times my seat belt has kept me from a totally unprofessional display."

"You let it roll off your back and keep going. You wait for the compliments, and the passengers who tell you how glad they are to see women in the cockpit, or what a nice flight they had; when flight attendants say you can land better than anyone they've flown with; when the captain tells you to watch out, there are going to be guys out there who are jealous of the good job you're doing. Or the day when your Mom, who is afraid of flying, doesn't flinch when you ask if she'd like to go up in the two-seat plane you just bought; with one engine and needing a paint job. And she is the same woman who cried when you told her you had soloed 'You haven't had enough time to go up alone, you're too young!' she said. "That makes it all worth while."

Condon married an airshow pilot and says they understand each other because both fly. "We don't have any problems in that area," she said, "because I don't really like to go upside down and although he thinks what I do is neat he has no need to do it himself."

Pam Noeldner (7/02/86 Piedmont Airlines)

Pam Noeldner earned her private license at seventeen while she was still in high school. She went to the University of Minnesota, continued getting her ratings, and taught for two and a half years. She took her Air Transport Pilot check ride two months before she turned 23. Then she flew canceled checks and freight for four months in Richmond, VA. In January 1985, she landed a job with a commuter airline. She worked there for eighteen months, eventually becoming captain on a Fairchild Metroliner. Piedmont hired her in July, 1986.

Noeldner said she had many good people guiding her in her career. Her dad flies privately, and it was through him that she got interested in flying. The Chief Flight instructor at the University of Minnesota, Linda Davis, was also a strong influence and she showed Noeldner unfailing support through the years. Her husband, also a pilot, gave her encouragement. Noeldner doesn't believe there is any inherent differences between male and female pilots as far as skills and learning curves go. "I have taught many girls and guys, and I think its just social pressure that causes the public to think of female pilots as unusual."

Gay King (03/02/87 Piedmont Airlines)

"Looking back, I have a vague remembrance of always wanting to fly," said Gay King. "My dad was a fighter pilot, and I remember when I was small his friends would come over and tell war stories. But it was my brother Ogden who introduced me to the air. He's a private pilot and he took me up in his single engine Cessna." (Fig. 8-16)

King majored in French and elementary education in college, and thought she wanted to be a teacher. Her teaching job lasted four months before she became disillusioned. She found it underpaid and dull. "I wanted a more exciting life so I became a flight attendant with TWA."

At first the job was all she thought it would be, but it soon became

Fig. 8-16 One Of Gay King's desires is to go up into space. Here she is practicing weightlessness in a KC-135

routine. It was also hard on her physically as she was flying half a dozen trips to Europe each month. "After some thought, I decided I wanted to do something not only exciting, but challenging. When I asked myself what I most wanted to do, the answer was be a pilot. My dad is a chauvinist - he'll admit it. He said I was wasting my time and money. He soon became my best fan and strongest supporter."

By the summer of 1979, she received her private pilot's license. "I loved it! It was the most rewarding thing I'd ever done." The following summer she earned her commercial, instrument, and multi-engine ratings. Then came the ATP and flight engineer written exams. "It was an all consuming passion - I lived and breathed flying. I didn't go to formal ground school. I studied on my own, so I had no social life. It was a small price to pay. I didn't miss it. I knew it would pay off in the end."

King wanted to work for Piedmont Airlines. She had 1,800 hours but only 300 hours of it was multi-engine time. "I ran into some problems when I tried corporate flying. One company said, the president's wife wouldn't approve of a female pilot. Another said my pay as a flight attendant was better and they didn't want to hire me and take me away from a good salary. So I took a leave of absence from TWA and went to Air Midwest, a commuter. My co-workers were supportive, but some didn't believe I'd do it. The pilots were helpful and gave me moral support."

The pilots and customers at Air Midwest accepted her with little or no resentment. "Once we were up in the air, everything is very professional. Most of the captains are younger and seem to have gotten used to the women's movement. A pilot realizes that if you're there, you evidently had to do something right to get there. I think the general public is more accepting, too. Some are still in dark ages. They become upset with a woman in the cockpit. A flight attendant once had to reassure a man who was asking, 'is it going to be OK, is she qualified?' It's annoying when some people are that way. It really hurts when women are like that. Do they have so little respect or belief in themselves that they assume all other women are the same way and can't do anything? I once had a man walk away from the ticket counter because there was myself and a female captain." Consumed by her passion, less than a year later, King had enough time to apply to Piedmont.

"Getting here wasn't easy. It took time and money, but I'd do it all over again. I love flying. Up there I have a window seat with the ever-changing scenery. Nothing is quite the same. Everyday I have new experiences. I find myself thinking what if I didn't fly? My mind goes blank. I can't think of anything else."

Motivation

Like all commercial pilots, women go through the same rigorous flight exams and written exams every six months, but often stricter standards seem to apply in judging them. In reaction, many women seem to be even more determined to prove themselves. Every landing and takeoff has to be perfect.

With more women flying than ever before, some interesting statistics are developing. A study of the accident records of the NTSB shows women had proportionally fewer accidents than men in every category examined (excluded were airline and military records). Since eighty-one percent of the accidents were pilot error the findings are significant. When women did have accidents there were fewer injuries. Male pilots were five times more likely to cause death to themselves or a passenger.

The reason for these results is not obvious and requires a look at our social structure. Some human behaviorists say society socializes men to be reckless and women to be prudent. Women and men had about the same number of motor-skill related accidents but men had almost three times the judgement-error accidents. Men had more judgement-error accidents than women had accidents in total. The judgement-error accidents included fuel mismanagement, inadequate preflight preparation and the selection of inadequate terrain. Motor skill errors are unforeseeable and include improper recovery in a bouncing landing or failure to maintain directional control.

The women have another factor on their side beside careful preparation and prudent judgement. They are more motivated to continue training. Women advance their certification at a greater rate than men.

Fig. 8-17 Some of the ISA Charter members (L-R, T) Angela Masson, Maggie Rose Sandy Donnolly, Jean Haley Harper, Holly Mullins, Beverley Bass, Karen Kahn. (B)Norah O'Neill, Gail Gorski Schlicht, Valerie Walker, Denise Blankinship, Terry Rinehart & Sharon Hilgers Krask

The bottom line is men have been more risk-oriented than women.

Times are changing but unfortunately some one forgot to tell a male passenger on a flight out of Seattle. One of Alaska Airlines' senior captains, Christi Gomes had just been pushed away from the gate. As she made the usual cabin announcement, a male passenger in his late 30s stood up and refused to allow the plane to go another inch. The other passengers tried to get him to calm down, but he would not fly on her flight. Captain Gomes decided to let him off. She had the back stairs lowered and the man slid into the shadows of the evening. Captain Gomes continued to her destination without further incident and the male passenger has not been seen since.

International Society of Women Airline Pilots - ISA + 21

In the fall of 1977, Stephanie Wallach, a pilot with Braniff Airlines, and Beverley Bass were dining together in New York City. They spoke about how much they would enjoy meeting the other women who were flying with U.S. airlines.

Fig. 8-18 The ISA at a recent convention

In January, 1978, Bass and Wallach were invited as representatives of their respective airlines to participate in a Zonta Club Program, honoring Amelia Earhart, in Washington, D.C. They used this opportunity to discuss their idea with five other pilots just to see if there would be a genuine interest in planning an informal get-together.

They returned home to New York and drafted a letter to the chief pilot at each airline employing female pilots. They had fears of not hearing from anyone, but that was not the case. In a short time people starting responding.

Before sending out the letter they had spoken with Claudia Jones, (then a First Officer with Southwest Airlines) and she graciously offered to have the convention in Las Vegas, where she was living at the time. Continental was the host airline, and Claudia, a past president of the Whirly-Girls, was responsible for all hotel and convention arrangements. (Fig. 8-17)

The two women had a dream. In May 1978, twenty one women pilots from ten U.S. airlines met in Las Vegas to share their common professional interests. They discussed various names and tried to come up with something catchy. ISA is a meteorological term used for an altitude pressure conversion and when they realized that the letters could also stand for "International Society of Women Airline Pilots," ISA + 21 was born. The women chose to form a social rather than a political organization, feeling that their respective pilot unions provided adequate representation of their needs. Growing each year, ISA + 21 today numbers over 400 members worldwide from thirteen foreign and twenty-two U.S. air carriers. There is also an auxiliary for the husbands. Husbands of Airline Pilots, "HALP." They wear "T" shirts with a drawing of a lady pilot dragging a man along by a white scarf."

The International Society of Women Airline Pilots links a wide range of aviation backgrounds and experiences and encourages it's members to help other women entering the profession. It maintains an Information Bank, networking, and service projects.

Each year they hold a convention and one highlight is the Captain's Club presentations, giving special recognition to ISA members who have completed their Left Seat check ride during the previous year. A mini-gathering is also held early each year to renew friendships, to meet new members, to express views, to relax, and have fun. (Fig. 8-18)

ISA's Information Bank is a network of women helping other women with the unique problems and struggles encountered in pursuing an airline pilot career. Unlike the pioneering women of the last half-century, the ISA now has people to assist and guide interested women. Questions received by the Information Bank are be referred to one or more women pilots in the writer's local area.

Each year they award scholarships to deserving young women, allowing them to pursue careers in aviation.

Fig. 8-19 Rita Reo has been flying for almost 19 years. She has flown in two air races, and used to be a Boeing 747 Flight Engineer for UPS. She is pictured here in a Fokker F-28.

Fig. 8-20. A few of the pilots who fly for Federal Express (L-R) Angela Allen, Florence Sanders, MegAnn Streeter, & Holly Mullins. The little figure on the right is future airline pilot Jeffrey Michael Mullins

In Memorial

On March 3, 1991, United Airlines Flight 585 was 3.3 nautical miles from Colorado Springs Municipal Airport, on final approach. Suddenly the Boeing 737-200 aircraft made an apparently uncontrolled roll to the right in a split "S" maneuver. The plane continued to roll about 180 degrees in its 6-9 second plunge vertically into the ground from 1,100 feet. Killed were Captain Harold Green, an experienced captain with 9,902 hours and First Officer Patricia K. Eidson, also an experienced copilot with 3,903 hours. Twenty passengers and two flight attendants were also lost. Patricia Eidson was the first American woman airline pilot to lose her life on duty.

Chapter 9

Aviation Entrepreneurs

"In the early days they said I was trying to make a statement, but I was just trying to make a living.

Capt. Bonnie Tiburzi

Airplane Sales

On May 20, 1927 Charles Lindbergh made his historic flight. Fifty years later that same spirit of meeting a challenge was alive in Gaye Gravely. It was a day that may have found its way into the dark recesses of her memory in spite of the fact that it was a memorable event in her life. On May 20, 1977 Gaye soloed, and the day became forever imprinted in her memory when her instructor told her that it was the 50th anniversary of Lindbergh's flight.

Gaye learned to fly, inspired by her father who was an ex-Army Air Force pilot. He founded the Veteran's Air Express after the World War II. At 23 Gravely was the youngest air carrier president in the business, with 104 people on the payroll and a fleet of DC-3s and C-54s flying passengers and cargo.Her dad named one of his DC-3 after her and perhaps that cemented her future ties to aviation. (Fig. 9-1)

He taught Gaye ground school when she was a freshman in high school, but she didn't begin flying lessons until she was thirty years old. Gravely said, "I always knew I'd be a pilot someday. Finally I made it

Fig. 9-1 Gaye Gravely

happen. I'd saved $1,700 but I was still worried that I'd run out of money before I got my private license. One Saturday in March, I took a "Discovery Flight." The next day I took my first lesson and on Christmas Eve, as a present to myself, I took my private check ride.

Gravely, who has a bachelor's degree in marketing from Arizona State University, had no idea of making aviation a career at the time, but that's what eventually happened. First in non-flying jobs, she served as director of business development for Aero Services, a national FBO chain, and after that she was the first woman supervisor of customer support for Garrett Engine, a division of Allied-Signal Aerospace, on Long Island, New York.

According to Gaye, today she has, "the ultimate job." She sells new airplanes for Beechcraft working out of the company's office at First Aviation, in Teterboro Airport, New Jersey. She is the first woman sales representative Beechcraft has had in more than a decade, and only their second one in their 59-year history.

"I was determined. I wanted to sell for Beach more than anything." she said. "I called and called and called but each time there were no openings. I continued to call and one day after almost a year of calling I called again and was told, 'Funny you should call today, someone just quit.'"

Gravely holds a commercial license with instrument and multi-engine ratings, all of which she's proud to say she paid for herself. But she had never sold aircraft before her job at Beech. "The vice president of sales told me, 'You've got the joy and the drive. I can teach you the rest.'" Gaye was getting in a lot deeper than she realized at the time. The fact that Beech produces 11 models from single-engine piston models through turbo-prop commuters and Beech jets didn't trouble her. Just learning all the performance features was a major chore. It was a challenge that excited her. She met the challenge head-on and daily continues to learn about the technical side of the airplanes and at the same time, the complexities of airplane sales.

"Selling an airplane is not at all like selling a car," Gravely points out. "You can walk into a car showroom and pick out a color and model based perhaps on gas milage or family needs. With an airplane it's more complicated and longer. The buyer has to consider many factors; tax advantage, utility, cost versus time saved on a business trip, and ancillary expenses like the savings on hotel rooms and meals. It's a long sell, and sometimes full of frustration," she says, but her enthusiasm sustains her during the long haul.

Flight Time

What do you do if an alumni group wants to charter a plane to take them to the superbowl? How do you handle the corporate client who needs to get two dozen top officers to meetings in six small cities in a week?

Traditionally the travel group gets on the phone and hopes against hope to round up a charter plane to take the group where they want to go as inexpensively and as quickly as possible.

That was the way things were done until three woman got together and came up with a simple and easier way.

Fig. 9-2 (L-R) Flight Time entrepreneurs Dara Zapata,
Jane McBride, & Patricia Zinkowski

Flight Time, an air charter clearing house, matches people to airplanes, saving travel agents and corporations time, work, and money. Dara Zapata, Jane McBride, and Patricia Zinkowski, all travel industry veterans, established Flight Time in 1985, and in their first year, they had over $1 million in bookings.Fig. (9-2)

The idea behind Flight Time is simple. There are over 12,500 airports around the country and only about 400 are serviced by scheduled airlines. The three women saw the potential to enable corporate America to reach a lot of destinations beyond the beaten track. They also knew that there were at least 21,000 airplanes of all makes and sizes sitting every day without a thing to do. We know that the only way airplanes make money is when they are flying, so if Flight Time could join up the corporate demand with the surplus airplane supply everybody could make money. Zapata, president of Flight Time, said deregulation had a lot to do with Flight Time's success. "Not as many airports are served regularly as there used to be, and industry is locating more plants outside major cities, and the people who have to travel to them can't get nonstop flights. We can get them where they want to go, conveniently and reasonably." For example, an agency representing the American Soybean Association had a dilemma. Nineteen officials needed to travel to six states in five days. Jane McBride, a private pilot, and CEO, said, "Flight Time arranged to charter a DC-3 for the group. The final cost was only $600 a head, right within the group's budget."

Zinkowski, chairman and chief financial officer added that Flight Time will arrange flights by matching customer requests in their data base with another data base listing over 1,000 aircraft ranging in size from a helicopter to an L-1011.

Flight Time has also grown to the point where they are looking for suitable sites for expansion.

Flight Time is an example of niche marketing and where women can achieve success in short order. The three entrepreneurs of Flight Time had an idea that centered around the ever-changing and continually-growing aviation industry, and took advantage of the opportunity.

Kim Darst

Whirly-Girl Kim Darst was planning a career in Marine Biology but that changed in November 1986, when her parents gave her a graduation present; a flight over the Grand Canyon. During the flight Kim watched the pilot, not the scenery. She was fascinated with what he was doing. When she got back to school she excitedly told her guidance counselor about the flight. Her counselor said her husband flew a helicopter and perhaps Kim would like a ride. Kim jumped at the chance, and hasn't come down since. It was "love at first flight." All thoughts of a career in Marine Biology slid quietly into the sea. By March 1987, Kim had her private helicopter license. She wasted no time putting it to good use. With the blessing of local police, school officials and the FAA, Kim helicoptered into her graduation ceremonies, and surprised most of her classmates. By November, 1987, one year after she had taken her first plane ride, Kim was a Certified Flight Instructor. (Fig. 9-3)

Kim also had total support from her parents along the way. Her mother drove her back and forth to Ellenville, N.Y., two-and-a-half hours each way for her lessons. Soon Kim realized commuting to get her ratings was too slow so she moved into a trailer near the airport. Then it was easy to spend 18-hours-a-day around airplanes. When she wasn't flying, she was learning the mechanic's end of the airplane or wiping snow

Fig. 9-3 Kim Darst

from the airplanes in bad weather. Today Kim has her single engine land, sea, commercial helicopter, and multi-engine ratings as well as her ground and flight instructor in both fixed wing, rotorcraft, and an Airframe and Powerplant license. In August, 1991, Kim became the youngest FAA examiner in the country.

One year after Kim earned her private license, she opened KD Helicopters in Blairstown, N.J. But before she could open her business officially, she needed a helicopter. Kim found just the one she was looking for, a Bell-47-G2, near Homestead, Florida. The problem: fly it back to Blairstown, New Jersey. On a warm sunny morning, in March with Ernie Kittner, her former flight instructor along, she took off from Florida. There were only minor problems along the way. The helicopter had no compass or heater. They overcame the first by following the "concrete compass" (Inter-state 95), and the sectionals. The second became more serious as they headed north. By the time they reached South Carolina they were switching control every fifteen minutes so each could warm up their hands. In Annapolis, Maryland, they were grounded by an ice storm, but just overnight. The next morning they found the rotor tie-down straps frozen to the rotor blades. They used a hair dryer to thaw the straps and were soon airborne. The trip only took 17 hours with eight stops for fuel.

Since then, KD Helicopters has blossomed. Kim is kept busy seven days a week. She flies pipeline patrol three days a week out of Pittsburgh, Buffalo, and Syracuse. The rest of the time she gives flight instruction, does spraying, and takes photographers on aerial photo expeditions, so far accumulating more than 3,800 hours Total Time. When weather has her grounded she does maintenance and works on her Instrument-Instructor rating.

To get some idea of Kim's success and excellent reputation she has one student who drives from Brooklyn, New York, and another drives from Wurtsboro, New York, each traveling two-and-a-half hours each way. Another flies up from southern New Jersey in his Cessna to take helicopter lessons with Kim. They must have heard about her record. In three years Kim has taught 18 students to fly, and each has passed their flight test on the first attempt.

Kim believes firmly in the Whirly-Girl philosophy of helping to promote women in helicopter aviation. One of her former students, Pat Labagh, is now a Whirly-Girl, too.What's in Kim's future? Someday she'd like to own a Jetranger (she already has 50 hours in one), and expand her business. Her advice to young women thinking of a career in aviation? "Go for it! Join the military. There you'll get top-notch flight instruction in sophisticated aircraft. The way I did it was the hard and expensive way, but it was worth it."

Betty Pfister

Betty Pfister has been quietly making a legend in aviation for more than 50 years. Her experience as a pilot is as distinguished as her work in the aviation community. She has qualified in more than 25 aircraft, from the Piper Cub through the P-39 (Aircobra) and the four-engine B-24 Liberator. She became Whirly-Girl #52 in 1963, became a glider pilot in 1966 and got her balloon rating in 1975. Beside being instrument and commercial rated in fixed wing and helicopters, she is considered one of America's outstanding race pilots. She is also a member of the US Precision Helicopter Team and competed in two World Championships.

Betty's aviation experience is not limited to civilian flying. She completed her WASP training in Sweetwater, Texas, in 1943, and served as a Ferry Command Pilot. Over the years she maintained her association with her war-time friends as a member of the Order of Fifinella, an ex-WASP organization. She also is a member of the Women's Military Association. (Fig. 9-4)

Betty had a wide range of aviation experiences after the war too. She worked as an instructor pilot, a DC-3 co-pilot for several non-scheduled airlines, an FAA accident prevention specialist, and as an assistant engineer in the Automatic Pilot Laboratory, of Bendix Aviation.

Betty has also held several judgeships in aviation. Beside being an International Judge for sport aviation competitions conducted under the aegis of the Federation Aeronautique Internationale, she has served as a judge during the 6th World Helicopter Championships in 1989. She is planning on being a judge in the 1992 National Helicopter Championships in Las Vegas and is one of the ten United States Judges authorized to participate in the 7th World Helicopter Championships in England.

Betty has also served on the FAA's Womens Advisory Committee on Aviation, and has been a member of the Ninety-Nines for more than 35 years, serving as its president from 1985 to 1987. She is the founder of the Helicopter Club of America and currently its Vice President.

Over the years, Betty Pfister has had friends like Jacqueline Cochran, Gus Grissom and Chuck Yeager, but perhaps her most meaningful achievements have resulted from her distinguished leadership in civilian

Fig. 9-4 Betty Pfister

aviation. She has served as a member of the Aspen Colorado Airport Authority and for the past 35 years has contributed to its growth from a seldom-used grass strip to a busy commercial airport. It was her inspiration that moved FAA officials to install a tower in 1969, even though Aspen did not meet the FAA criteria. Betty also planned and supervised the construction of Aspen Valley Hospital Heliport in 1966, the first in Colorado. She went on to manage the facility for many years. Along with the late Senator Peter Dominick, Betty was responsible for the enactment of the law that required aircraft to be equipped with an emergency transmitter.

One of the highlights of her career has been the Pitkin County Air Rescue Group which she founded in 1968. Today she still serves as its chairman. This unique organization consists of 20 of the best pilots in the Aspen area and takes the responsibility from the Civil Air Patrol in search for overdue and downed aircraft in a 50-mile radius of Aspen. There is no "red tape" in this group. They provide quick response by pilots familiar with the local terrain. Over the past 22 years they have been instrumental in saving the lives of 26 people. Betty also organized the Pitkin County Helicopter Rescue Fund, and raised enough capital so the interest sustains emergency helicopter rental expenditures regardless of the victim's ability to pay.

For her life-long contributions to aviation, Betty was inducted into the Colorado Hall of Fame in 1984. In addition to all these remarkable years of selfless giving, Betty and her husband Art have raised three daughters.

Evelyn Bryan Johnson

They call her "Mama Bird" because she makes sure her fledgling air pilots return to the nest safely. Evelyn Johnson has been living and breathing aviation for more than 47 years. Her love for the sky began in 1944, when her husband was off in the war. Bored in his absence, she didn't want to take up knitting, sewing, cooking, or any other typical "keep busy" activity. One day while heading for work in her dry cleaning store, she noticed a sign "Learn to Fly!" That sounded exciting, she thought. Why not? Evelyn is still flying today, seven days a week, 10-12 hours a day. She shrugs off the fact that she turned 81 in November 1990, and she doesn't dwell on the fact that she passed the 50,000 hour mark on May 7, 1991. "I'm working on my second 50,000 hours," she said optimistically, estimating she'll be "around 130 years old" when she reaches that mark. To put her achievement in perspective, 50,000 hours is equivalent to spending 24-hours a day in the air, for 5.7 years.

Evelyn is currently an FAA examiner and flight instructor and has lost track of all the students she has taught how to fly. She estimates proudly that about 40 percent of her students have gone on to commercial aviation jobs, many with the airlines. She does know that she has given more than 9,000 check rides, and those who have been privileged to fly with her know she is no soft touch. About 26 percent of her students flunk on their first ride. "If they can't navigate by reading a sectional and recognizing landmarks, they get a pink slip," she says. Evelyn's emphasis is on pilotage and seat of the pants flying. (Fig. 9-5)

Fig. 9-5 Evelyn Johnson

Evelyn has several career highlights to be proud of too. In her 47 years of flying she has never scratched an airplane or student, even though she has had two engine failures and an in-flight fire.

In 1958, shortly after she earned her helicopter rating (she is Whirly-Girl #20, and the 4th woman in the world to get a helicopter instructor rating) she saved a pilot from a crashed helicopter. The crash occurred at the airport she managed and she was the closest to the scene. Shortly after taking on fuel a helicopter rose into the sky. Moments later it came spiraling down and crashed on the field. The crash killed the passenger outright, pinned the pilot in the wreckage and split the fuel tank open. Without hesitating Evelyn grabbed a fire extinguisher and rushed to the machine. The rotor was still turning and the fuel was leaking. She splashed through the fuel, and crawled on her knees to reach the cockpit. She shut off the ignition switch, and foamed down the overheated engine. Then she looked at the pilot. He had a back injury she thought, and her first aid training told her it was best not to move him. She was right. The pilot had a broken back and moving him would have killed him. For her bravery she won the Carnegie Medal. The pilot lived and went on to spend twenty five years as an FAA test pilot.

How long does Evelyn intend to keep flying? "Just as long as I can pass the FAA's First Class physical. That's the same one they give airline pilots," she says beaming.

The Last Word

Over the last eight decades women have fought an uphill battle for acceptance, recognition and jobs. They persevered and as a result, there have been vast strides forward in women's ability to gain employment in aviation. The number of women in non-pilot activities nearly doubled in the 1980s, and the number of women airline mechanics more than tripled. In 1970, there were no female Flight Engineers. By 1989, there were 1,042. Women account for almost six percent of airline pilots and the number of women with commercial licenses rose more than 20 percent between 1981 and 1989.

The outlook for women in the decade of the Nineties is brighter than ever. The United States is suffering from a pilot shortage caused by the mandatory retirement of pilots who were trained in wartime and entered civilian aviation. Fewer pilots are leaving the military for civilian jobs and business and pleasure flying are also increasing the pilot demand. In the next decade the airlines are expected to hire between 52,000 and 62,000 pilots to meet this shortage. Airlines and corporations are actively recruiting qualified pilots regardless of sex, at competitive salaries, and women's salaries in general aviation are consistent with their male counterparts.

Many of the women featured in this book have noteworthy accomplishments based on the prefix "first female," and there are still opportunities and milestones for today's generation. But the prefix "first female" is not the only thing that can be said about them. The "first" and those that followed are role models for this and future generations of women aviation professionals. Today's role models will be the heroines to tomorrow's generation.

Their commitment to performance as a chief measurement of success has proven to be the key to opening the door for other women who love to fly. Their significance today is more than the sum of their individual accomplishments. The Ninety-Nines, the International Society of Women Airline Pilots, and the Whirly-Girls, have formed strong and cohesive networks. The strength of their groups has allowed the individuals to move out of the group and be integrated into the mainstream of aviation.

Women now have access to the education, legal standing and the professional networks that will allow them to walk through the doors opened by the women in the 1970s and 1980s, and into a new era of full participation and equality in aviation.

Following Amelia Earhart's tragic disappearance in 1937, columnist Walter Lippman said he could not remember the stated purpose of Miss Earhart's last flight. But more important, he said, "was the spirit that animated the venture that counted for far more than any practical results in the development of aviation." That same spirit of forging ahead, breaking barriers, and establishing new records is still alive in today's "Ladybirds."

Bibliography

Books
1- Bilstein, Roger E. *Flight in America.* Baltmore, Md: Johns Hopkins University Press.
*2- Briggs, Carole S. *Women in Space - Reaching the Last Frontier.* Minneapolis Learner Publ. Co. ,1988.
*3- Cochran, Jacqueline; Brinley, Maryann Bucknum: *Jackie Cochran:The Autobiography of the Greatest Woman Pilot in Aviation History.* New York: Bantam Books., 1987.
*4- Douglas, Deborah. *United States Women in Aviation 1940-1985.* Washington, D.C. Smithsonian Studies in Air and Space. #7. 1990
*5- Earhart, Amelia. *The Fun of It.* New York: Harcourt Brace; 1932.
6- Goldman, Nancy,ed. *Female Soldiers: Combatants or Noncombatants.*"Historical and Contemporary Perspectives:" Greewald Press, *1982.*
7- Hamlen, Joseph. *Flight Fever.* New York: Doubleday & Company, 1971
8- Harris, Sherwood. *The First to Fly. Aviation's Pioneer Days.* New York; Simon & Schuster; 1970.
9- Hodgeman, Ann; Djabbaroff, Rudy. *Skystar History of Women in Aviation.* New York. Atheneum, 1981.
10- Holm, Jeanne. *Women in the Military: An Unfinished Revolution.* Novato, California: Presidio Press, 1982.
11- Keeton, Kathy. *Women of Tomorrow.* New York St. Martins/Marek Press; 1985.
12- Leuthner, Stuart ; Jensen, Oliver. *High Honor .* Washington, D.C. Smithsonian Press; 1989
13- Lindbergh, Anne Morrow. *North to the Orient.* New York. Harcourt Brace & World. 1935.
*14- Lomax, Judy. *Women of the Air.* New York, Dodd; Mead, & Company; 1987.
15- May, Charles Paul. *Women in Aeronautics.* New York: Thomas Nelson &Sons, Inc.,1962.
16- Moolman, Valerie. *Women Aloft.* Virginia: Time Life Books 1982.
*17- Oakes, Claudia. *United States Women in Aviation Through World War I.* Washington, D.C. Smithsonian Studies in Air and Space #2, 1979.
*18-_____*United States Women in Aviation 1930-1939.* Washington, D.C. Smithsonian Studies in Air and Space #6, 1985.
*19- Pazmany, Kathleen Brooks. *United States Women in Aviation 1919-1929.* Washington, D.C. Smithsonian Studies in Air & Space #5,1983.
20- Peckham, Betty. *Women in Aviation.* New York:Thomas Nelson & Sons, Inc. 1945.
21- Planck, Charles E. *Women With Wings.* New York: Harper Brothers 1942.
*22-Ricks, Chip. *Beyond the Clouds.* Wheaton, Il.: Tyndale House Publishers, Inc.1979. **
23- Thaden, Louise M. *High Wide and Frightened.* New York. Stackpole & Sons.,1938.
*24- Tiburzi, Bonnie. *Takeoff.* New York: Crown Publishers, 1984.

*25- Van Wagenen-Keil, Sally. *Those Wonderful Women in Their Flying Machines.* New York. Rawson, Wade Publishers, Inc. 1979.

26- Villard, Henry Serrano. *Contact :The Story of the Early Birds.* New York. Thomas Y. Crowell, 1968.

27- Wertheimer, Barbara Mayer. *We Were There. The Story of Working Women in America.* New York. Pantheon Books, 1977.

28- Wood, Winifred. *We Were WASPs* . Coral Gables,Glade House,1945.

29- Wilson, Andrew. *Space Shuttle Story.* New York: Cresent Books; 1986.

30- *The History of the Ninty-Nines.* Oklahoma. The 99, Inc, Publishers. International Organization of Women's Pilots; 1979.

31- Hopkins, J.C.; Goldberg, Sheldon, A. *Development of the Strategic Air Command 1946-1986.* Office of the Historian, Headquarters Strategic Air Command, Offut Air Force Base, Nebraska; Sept 1986;
* Recommended reading

Magazines

1- Baroni, Diane. "Flying High With Sally Ride." *Cosmopolitan.* September, 1984.

2- Bateman, Major, Sandra. *Air Power Journal* "The Right Stuff Has No Gender." Winter 1988.

3- Brown, Margery. "What Men Flyers Thing of Women Pilots." *Popular Aviation & Aeronautics March 4, 1929.*

4- Buffington, H. Glenn. "Flair for Flight" *99 News* . M/A 1974.

5- Burkett, Katherine. *Ms.* "Flying the Safer Skies." June 1987.

6- Burnett, Gene. "A Florida Native Born to Fly." *Florida Living* ;Sept. 1977.

7- Chadwick, Bruce. "Women in Aviation: They Make Winging It Pay." *Cosmopolitan.*November, 1983.

8- Christmann, Timothy J. "Navy's First Female Test Pilot." *Naval Aviation News.* Oct. 12, 1986.

9- Chun, Victor "The Origin of the W.A.S.P.s." *American Aviation Historical Society Journal.* Vol. 14 #4 Winter 1969.

10- Collins, Helen. "From Plane Captains to Pilots." *Naval Aviation News.* July 1977.

11- Conant, Jennet, et al., "Women in Combat?" *Newsweek.* Nov. 11, 1985.

12- Gray, Mellisa. "Women to the Front." *Flying.* September 1980.

13- Grossman, Ellie."Soaring High: Careers in Aviation." *Working Woman.* May 1980.

14- Hill, Carol. "Hero For Our Time." Vogue. August, 1983.

15- Howard, Jean Ross. "Our Egg Beaters Aren't in the Kitchen." *Rotor Wing.* May 1991

16-   "Whirly Girls Prove That Helicoptering Isn't Only for Men."*Rotor Wing.* October 1967;

17- Kaplan, Mary-Jo. "Women in Nontraditional Jobs." *Cosmopolitan.* June, 1983.

18- Kerfoot, Glenn. "Propeller Annie."*Aviation Quarterly.* Vol. 5 #4 1979;

19- Khalil, Abdullah; Wells, Grady; "A Galaxy of Expectations." *Sisters.* Spring 1989.

20- Laboda, Amy. "Woman With Navy Wings." *Flying.* January 1990.

21- Larsen, Will; Kaden, G.L. "Women With Navy Wings." *All Hands.* April, 1975.

22- Melling, Lori. "I Am A Navy Pilot." *Cosmopolitan.* January ,1991.
23- Moses, Sam. "Sky Princess Passes On Her Secret." *Sports Illustrated.* 12/18/89.
24- Orr, Verne. "Finishing the Firsts." *Air Force Magzine* . Feb.1985.
25- Poole, B.E. "Requium for the WASP." *Flying.* Dec. 1944.
26- Powell, Hugh. "Harriet Quimby America's First Woman Pilot." *American Aviation Historical Society Journal.* Winter 1982.
27- Spencer, Scott. "Epic Flight of Judith Resnick." *Esquire.* Dec. 1986.
28- Stine, G. Harry. "Women Drivers." *Analog Science Fiction/Fact.* 1/89.
29- Strother, Lt. Col. Dora Dougherty, USAFR. "The W.A.S.P. Training Program." *American Aviation Historical Society Journal.* Winter 1974
30- Rucker, Cindy. "Struggle for Stripes." *Air Progress.* Dec. 1979.
31- Russell, Sandy. "High Flying Ladies." *Naval Aviation News.* February 1981.
32- Samuelson, Nancy B. Lt. Col. USAF. "Equality in the Cockpit." *Air University Review.* May-June 1984.
33- Stuart, J. "The Wasp." *Flying,* Jan. 1944.
34- Weston George. The Beauty and the Bleriot"*Aviation Quarterly.* "Vol. 6 #1 1980.
35- Congressional Research Services. "Women in the Armed Forces." CRS-9 Library of Congress Washington D.C. 9/25/85.
36- *Good Housekeeping.* Sept. 1912.
37- *Parade.* July 19, 1987.
38- *Time.* "All in the Family."March 24, 1980;
39- *Ebony.* "Show Me Girl Shows the Way" April, 1980.
40- *Ebony.* "They Take to the Skies." May 1981.
41- *Ms..* September 1986; "A Woman's Place is in the Sky."
42- *Vogue.* January 1984.
43- *Naval Aviation News.* Obituary Lt. Cmdr. Rainey.October 1982.
44- *Naval Aviation News.* "Records" November 1981.
45- *Seventeen.* February 1981; p.71.
46- *Working Woman.* "Those Daring Woman in their Flying Machines." May 1981.
47- Women Pilots in the Army Air Force. 1941-1944. AA Historical Studies 1945.
48- *Propwash* Winter 1988.
49- *GEO* September 1982 p.108.
50- *Life* Magazine July 19, 1943.
51- *Look* Magazine Feb. 9, 1943.
52- *Flying* ." Rerun: 1929 Women's Air Derby" *August 1989.*

Newspapers
1-*New York Times.* Various issues
2-*N.Y. Daily News.* 7/23/87 "Just Plane Great." Sherryl Connolly.
3-*Chicago Tribune.* Oct. 29, 1986 "Fly Right Past the Right Stuff."Bob Greene.
4-*Winston-Salem Journal.* October 21, 1986 "She's Flying High."
5-*Christian Science Monitor.* April 6, 2982 "Whirly-Girls: Aloft For a Living." Diane Casselberry Manual.
6-*Christian Science Monitor.* November 28, 1979 "Amelia Would be-Proud." Randy Shipp.
7-*Stapleton Interline.* 2/5/88 "Woman Marks Milestone In Aviation" Scott E. Dial p.3.

8- *Beacon.* Bill Reed."Top Flight" April 29, 1986.
9- *Boston Globe* July 2, 1912
10- Newark Star Ledger, "Wings of New Jersey." Jack Elliott, 12/30/90
11- *Washington Times*, 4/29/85 p.3B.

Other
1- *Elements Of Organizational Discrimination: The Air Force Response to Women As Military Pilots* Dissertation by Angela Masson University of Southern California, January, 1976.
2- Various NASM biographical files
3- Survey of American Women Helicopter Pilots; Enid C. Kasper, 1988.
4- Whirly Girls News Release. 3/4/63
5- ICAO Bulletin 12/75. "The Whirly Girls: International Flying Ambas sadors."
6- Air Force Fact Sheet 87-45 Nov. 1987
7- Air Force Fact Sheet 35-60 Jan. 20, 1986

Contributors
Listed alphabetically with rank or title at the time of last correspondence.
Colleen Andersen, Lt. Jg. USCG.
Cynthia Axel, Lt. Jg. USCG.
Beverley Bass, Captain, American Airlines.
Denise Blankinship, Captain, USAir
Janis Blackburn, Second Officer, Eastern Airlines.
Christopher J. Canfield, Captain USAF.
Barbara Collins, Flight Officer - Washington, DC -PD
Amy Correll, Captain USAir.
Camela Jo Condon, First Officer, USAir.
Cheri O'Donnell-Curley, Captain, Suburban Airlines.
Kim Darst, KD Helicopters
Gail A. Donnolley, Lt. USCG.
Connie Engel, Major, USAF.
Denna Gollner, First Officer, United Airlines.
Gaye Gravely, Beachcraft Sales Representative
Lori L. Griffith, Captain, USAir.
Laura H. Guth, Lt. Jg. USCG
Leigh Herrmann, KTAR Radio, Arizona
Terre W. Hines, First Officer, USAir.
Jean R. Howard, Chairman of the Board, Whirly-Girls Inc.
Joseph P. Hopkins, United Airlines.
Charmienne S. Hughes, President, Triad Aviation
Evelyn Bryan Johnson, FAA Examiner
Melissa Keiser, Librarian, NASM.
Gay King, First Officer, USAir .
Les A. Kodlick, 1st Lt. USAF.
Rosemary Mariner, Lt. Cmdr. USN.
Holly Mullins, Captain, Federal Express
Pam Noeldner, First Officer USAir.
Kerry L. Matheison, Sgt. USAF.
Pamela McFadden, Consultant.
Teresa McIntosh, Flight Officer, LAPD
Angela Masson, Captain, American Airlines.
Joan Morris, Photo Archivst

Norah O'Neill, First Officer, Flying Tigers.
Jody Norman, Photo Archivst
G. Putnam, Major General USAF (Ret.)
Rita R. Reo, First Officer, USAir.
Duana M. Robinson, First Officer, USAir.
Sally Roever, First Officer, Midway Airlines.
Carole Sandhovel, Research Assistant
Alda L. Siebrands Lt. Jg. U.S.C.G.
Werner Siems, USCG Office of Consumer Affairs.
Lennie Sorenson, Captain, Continental Airlines.
Joseph W. Steiner, Major USAF.
Pamela Mitchell Stephens, First Officer, Northwest Airlines.
Kathleen Sullivan - First Officer, USAir.
Bonnie Tiburzi, Captain, American Airlines
Jeannie Thompson - First Officer, Continental Airlines.
Jean Tinsley, President Helicopter Club of America
Elizabeth Uhrig, Lt.Jg. USCG.
Deborah Utz, First Officer, Piedmont Airlines.
Ms. Anna C. Urband, Dept. of The Navy.
Dorothy Vallee, First Officer, Piedmont Airlines.
Emily Warner, FAA Examiner
Claudia P. Wells, Lt. U.S.C.G.
Bonnie Wilkens
Larry Wilson, Librarian, NASM.
Marty Wyall, Historian, WASPs.
Lucy B. Young - Second Officer, USAir: Lt.Cmdr C-9 TAC.
Patricia Zinkowski, Flight Time

Photo Credits
Cover, Preface, Intro, Smithsonian Institution, Author's photo by Nancy Holden, Lori Griffith, courtesy of Lori Griffith. Smithsonian Institution: p. 16, 20,22,26,31,42,47,58,65,72,82,(4-5) 90,94, 96, 140; Florida State Archives p. 38, 54. HV Pat Reilly, Aviation Hall of Fame of N.J. p.60. Marty Wyall WASPs p. 76,80,82 (4-4) 84,86,89,90,92. Jean Ross Howard p.97. p.98 Kim Darst. p.98 Jean Tinsley. p.98 Teresa McIntosh p.102, Charmeinne Hughes p.102, Colleen Nevius, p. 105. Leigh Herrmann p.104.. USN p.111, 113,116, 120. P. 122. Connie Engel. p. 125, 127. Colleen Andersen. NASA p.144,146, 148. P. 154, Emily Warner, p.154 Bonnie Tiburzi, p.155 Angela Masson, p.158 Beverley Bass, p.159 Lennie Sorenson, p.162 Holly Mullins, p.165 Denis Blankinship, p.166 Duana Robinson, p.169, Dorothy Vallee, p.170, Pam Stephens p.171, Amy Correll, p.172, Lori Griffith, p.174, Janis Blackburn, p. 179, Kathleen Sullivan, p. 182 Jeannie Thompson, p. 184, Gay King, p.188, Holly Mullins, p. 190 & 193, ISA, p. 191, Rita Reo p.193, Gaye Gravely, p. 195 Patricia Zinkowski, p. 197 Kim Darst, p. 198, Betty Pfister p. 200, Evelyn Johnson p.202 .

H

I

J

K

L

M

N

O

Appendix
For further information on careers in aviation contact the following organizations:

ISA + 21
International Society of Women Airline Pilots
P.O. Box 38644
Denver, CO. 80238

Ninety-Nines, International Women's Pilots
Will Rogers Airport
PO Box 59965
Oklahoma City, OK. 73159
Loretta Jean Gragg, Executive Dir.

International Women's Helicopter Pilots Assoc. (Whirly-Girls)
7551 Callaghan Rd. Suite 330
San Antonino, TX. 78297
Colleen Nevius, Executive Director

Women Military Aviators
PO Box 396
Randolph Air Force Base, TX. 78148
Lt. Col. Kelly Hamilton-Barlow, Pres.

Women in Aerospace
6212-B Old Keene Mill Ct.
Springfield, VA. 22152

Future Aviation Professionals of America
4959 Massachusetts Ave.
Atlanta, GA. 30337

Order Form

Did You borrow this copy of *Ladybirds*?

If so then now is the time to order your own personal copy, or a copy for a friend!

Send $19.95 + $2.00 Shipping & Handling:

Black Hawk Publishing Company
Post Office Box 24
Mt. Freedom, N.J. 07970-0024

Make Check or Money Order to: **Black Hawk Publishing Company.**

Ship to: Name:_____

Address:_____

City:_____ State:_____Zip:_____

Gift section:

I would like the author to sign this copy to:

Name_____

This book is available to educational institutions at bulk discount rates. Write the publisher for more information.